Contents

Fully endorsed by Edexcel, this ***Oxford GCSE for Edexcel Higher Revision Guide*** contains all the material you need to help you prepare for your GCSE Higher tier examination. Each topic contains:

- keypoints, examples and exercises to help you revise and fully understand the learning objectives

- past Edexcel exam questions to give you essential exam practice.

A practice exam paper is provided at the back of the book, as well as full answers to all questions.

The accompanying CD-ROM contains extra resources to aid your revision including examiners' tips and exam specification matching grids.

- Rounding gives the approximate size of a number.

  32 524 rounded to the nearest 1000 is 33 000.

- Numbers can be rounded using **significant figures.**

  75 345 rounded to 2 significant figures is 75 000.

- The **upper** and **lower bounds** are the two limits that a number can take.

  If 1600 is a number rounded to the nearest 100, the upper bound is 1650 and the lower bound is 1550.

- A negative number is a number less than zero.

- An integer is a positive or negative whole number (including zero).

  ..., −3, −2, −1, 0, 1, 2, 3, ..., are the integers.

- A **factor** is a number that divides exactly into another number.

  1, 2, 3, 4, 6, 8, 12, 24 are all factors of 24.

- Common factors are factors that are shared by two or more numbers.

  5 is a common factor of 30 and 45.

- A **prime number** is a number with exactly two factors.

  17 is a prime number. The only factors of 17 are 1 and 17.

- A prime factor of a number is a prime number which is also a factor of the number.

  7 is a prime factor of 21.

- Any number greater than 1 can be written as a **product of prime factors**.

  $80 = 2 \times 2 \times 2 \times 2 \times 5 = 2^4 \times 5$

- You multiply a number by an integer to get a **multiple**.

  The multiples of 8 are also the 8 times table
  8, 16, 24, 32, 40, ...

- Common multiples are multiples that are shared by two or more numbers.

  24 is a common multiple of 6 and 8.

- The Highest Common Factor (HCF) of two numbers is the largest number that is a factor of both of them.

  The HCF of 16 and 20 is 4.

- The Least Common Multiple (LCM) of two numbers is the smallest number that is a multiple of both of them.

  The LCM of 24 and 40 is 120.

Keywords
Factor
Multiple
Prime number
Product of prime factors
Significant figures
Upper and lower bounds

Write the first two non-zero digits. 000 are not significant, but are used to make 75 000 the right size.

Factors come in pairs.
$1 \times 24 = 24$; $2 \times 12 = 24$
$3 \times 8 = 24$; $4 \times 6 = 24$

1 is not a prime as it has only one factor.

2 and 5 are prime numbers.

$6 \times 4 = 24$
$8 \times 3 = 24$

Factors of 16 : 1 2 4 8 16
Factors of 20 : 1 2 4 5 10 20

Multiples of 24 : 24 48 72 96 **120**
Multiples of 40 : 40 80 **120**

Example

Write these integers in order of size. Start with the smallest.

a 25 431   2099   28 990   2987   19 345

b 15   0   −4   −16   −21

TTh	Th	H	T	U
2	5	4	3	1
	2	0	9	9
2	8	9	9	0
	2	9	8	7
1	9	3	4	5

a 2099   2987   19 345   25 431   28 990

b −21   −16   −4   0   15

Find the upper and lower bounds of the rounded numbers.

a 250 has been rounded to the nearest 10

b 46 000 has been rounded to the nearest 1000

Halve the 'nearest number', then add and subtract this to the rounded number to find the bounds.

a Half of 10 is 5.   Upper bound = 250 + 5 = 255

Lower bound = 250 − 5 = 245

b Half of 1000 is 500.   Upper bound = 46 000 + 500 = 46 500

Lower bound = 46 000 − 500 = 45 500

Round 2457 to   **a** 3 significant figures

b 2 significant figures

c 1 significant figure.

3 sig figs means you are only allowed 3 digits, although you can add 0s to make the number the right size.

a 2457 to 3 sig figs is 2460

b 2457 to 2 sig figs is 2500

c 2457 to 1 sig fig is 2000

Write 60 as the product of prime factors.

Product means multiplication.

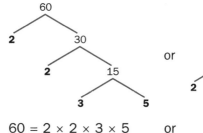
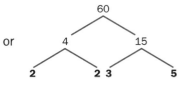

2	60
2	30
3	15
5	5
	1

$60 = 2 \times 2 \times 3 \times 5$ or $60 = 2^2 \times 3 \times 5$

See N5 for indices and powers.

HCF – Pick out the common factors and multiply together.

a Find the Highest Common Factor (HCF) of 45 and 60.

b Find the Least Common Multiple (LCM) of 45 and 60.

LCM – Calculate the HCF and multiply by the remaining factors.

a $45 = 3^2 \times 5$ **b** $45 = 3^2 \times 5$

   $= 3 \times \mathbf{3} \times \mathbf{5}$ $= 3 \times 3 \times 5$

 $60 = 2^2 \times 3 \times 5$ $60 = 2^2 \times 3 \times 5$

   $= 2 \times 2 \times \mathbf{3} \times \mathbf{5}$ $= 2 \times 2 \times 3 \times 5$

The HCF is $\mathbf{3} \times \mathbf{5} = 15$.   The LCM is $15 \times 2 \times 2 \times 3 = 180$

Exercise N1

1 Find all the factor pairs of 120.

(H p8, 11)

2 Two buoys ring to warn boats of some dangerous rocks. Buoy A rings every 45 seconds and Buoy B rings every 35 seconds. If the buoys start ringing at the same time, when will be the next time they ring together?

(H p10)

3 The attendance at a concert is 45 328. Round this number to

a 4 sig figs   **b** 3 sig figs   **c** 2 sig figs

(H p38, H+ p38)

4 The volume of a cuboid is 4199 cm³.

Each dimension of the cuboid is a prime number.

Find the length, width and height of the cuboid.

(H p8)

Volume of cuboid = $l \times w \times h$

5 The height of the mountain Ben Nevis in Scotland is 4400 feet, to the nearest 100 feet. Calculate the upper and lower bounds of the height.

(H+ p40)

6 Copy and complete the table. The first line is already complete.

	Number	Prime/Not prime	Reason
a	135	Not prime	More than two factors, such as 1, 5, 135
b	252		
c	483		
d	2401		
e	47		

(H p8)

7 When Liz was asked to estimate 60 seconds, she could only correctly guess the 60 seconds to the nearest 10 seconds.

a Calculate the upper and lower bounds of her estimate.

b Give the largest and smallest possible times if Liz estimates 5 minutes.

(H+ p40)

8 a Express each number as a product of prime factors.

    i 30   **ii** 75   **iii** 48

  b Find the Highest Common Factor (HCF) of these pairs of numbers.

    i 30 and 75   **ii** 30 and 96   **iii** 48 and 75

  (H p10, H+ p10)

9 a Express each number as a product of prime factors.

    i 36   **ii** 64   **iii** 80

  b Find the Least Common Multiple (LCM) of these pairs of numbers.

    i 36 and 64   **ii** 64 and 80   **iii** 36 and 80

  (H p10, H+ p10)

10 A hot air balloon can safely carry people weighing a total of 300 kg.

The weights of four people are 60 kg, 70 kg, 80 kg and 80 kg.

These weights are given to the nearest 10 kg.

Could the combined weights of the people be too heavy for the balloon?

Explain your answer.

(H+ p40)

11 By rounding all the numbers to one significant figure, write a calculation that estimates the answer for each of these calculations.

  a $\dfrac{38 \times 429}{53}$ **b** $\dfrac{215}{21 \times 109}$ **c** $\dfrac{45 \times 46}{45 + 46}$

(H p38, H+ p38)

12 a Express 315 as a product of prime factors.

  b What is the lowest number that 315 must be multiplied by, to become a square number?

  (H p10, H+ p10)

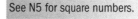

See N5 for square numbers.

13 The number 40 can be written as $2^m \times n$, where m and n are prime numbers.

Find the value of m and the value of n.

(*Edexcel Ltd., 2005*) 2 marks

14 a Use the information that

    $13 \times 17 = 221$

    to find the Lowest Common Multiple (LCM) of 39 and 17.

  b Use the information that $13 \times 17 = 221$ to write the value of

    i 1.3×1.7 **ii** $22.1 \div 1700$

See N4 for decimal calculations.

(*Edexcel Ltd., 2004*) 4 marks

Keywords
Decimal places
Equivalents
Irrational
Recurring decimal
Significant figures

- A fraction is a way of describing part of a whole.

  $\frac{4}{5}$ means 4 parts out of 5.

  $\frac{4}{5}$ ← numerator
  $\frac{4}{5}$ ← denominator

- A decimal is written using place values.

Tens 10	Units 1	.	Tenths $\frac{1}{10}$	Hundredths $\frac{1}{100}$	Thousandths $\frac{1}{1000}$
9	6	.	2	7	8

96.278 =

The digit 9 stands for 90

The digit 6 stands for 6

The digit 2 stands for $\frac{2}{10}$

The digit 7 stands for $\frac{7}{100}$

The digit 8 stands for $\frac{8}{1000}$

- A **recurring decimal** is a decimal that repeats itself forever.

  0.090 909 090 909 ... ... is a recurring decimal.

  $0.090\,909\,0... = 0.0\overset{..}{9}$

- A terminating decimal is a decimal with digits that eventually end.

  0.125 is a terminating decimal.

- A number is **irrational** if it cannot be written as a fraction. Irrational numbers do not have a recurring sequence of decimal digits.

  The surd $\sqrt{5}$ and π are irrational numbers.

  $\sqrt{5} = 2.236\,06...$
  $\pi = 3.141\,592...$

  See N5 for surds.

- A percentage (%) is a number of parts out of 100.

  $61\% = 61$ parts out of $100 = \frac{61}{100}$

- You can find equivalents for fractions, decimals and percentages.

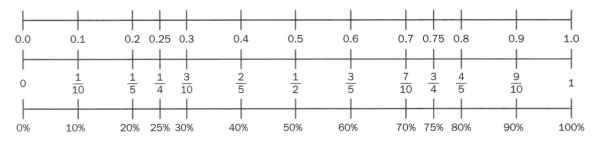

- You can order fractions, decimals and percentages using **equivalents**.

  0.445 $\frac{9}{20}$ 47.5% are in order of size, smallest first.

  $0.445 = 0.445$
  $\frac{9}{20} = 0.45$
  $47.5\% = 0.475$

- You can round numbers to their approximate size using

  - **decimal places** 8.755 rounded to 2 decimal places is 8.76

  - **significant figures** 8.054 rounded to 3 sig figs is 8.05

  Write 2 digits after the decimal point.
  Write 3 digits only.

Example

Suzy was given a mental arithmetic test of 40 questions.

She got 3 questions wrong.

Express her mark as a fraction.

She got 40 − 3 = 37 questions right, so her mark was $\frac{37}{40}$.

Example

a Change $\frac{3}{8}$ to   **i** a decimal   **ii** a percentage.

b Change 0.625 to **i** a percentage   **ii** a cancelled fraction.

c Change 47.5% to **i** a cancelled fraction **ii** a decimal.

a **i** $\frac{3}{8} = 3 \div 8 = 0.375$

   ii $0.375 = \frac{375}{1000} = \frac{37.5}{100} = 37.5\%$ **or** $\frac{3}{8} \times 100 = 37.5\%$

b **i** $0.625 = \frac{625}{1000} = \frac{62.5}{100} = 62.5\%$ **or** $0.625 \times 100 = 62.5\%$

   ii $0.625 = \frac{625}{1000} = \frac{25}{40} = \frac{5}{8}$

c **i** $47.5\% = \frac{47.5}{100} = \frac{475}{1000} = \frac{95}{200} = \frac{19}{40}$

   ii $47.5\% = \frac{475}{1000} = 0.475$

Example

Order these numbers, smallest first.

21% $\frac{3}{16}$ $\frac{1}{5}$ 0.19

21% = 0.21 $\frac{3}{16} = 0.1875$ $\frac{1}{5} = 0.2$ 0.19 = 0.19

$\frac{3}{16}$ 0.19 $\frac{1}{5}$ 21%

> Convert the numbers to the same type, either fraction, decimal or percentage – your choice.

Example

Express these **recurring decimals** as fractions.

a $0.\dot{8}$ **b** $0.\dot{3}\dot{6}$ **c** $0.58\dot{3}$

a $x = 0.\dot{8}$ **b** $x = 0.\dot{3}\dot{6}$ **c** $x = 0.58\dot{3}$

    $10x = 8.\dot{8}$ $100x = 36.\dot{3}\dot{6}$ $3x = 1.75$

     $9x = 8$ $99x = 36$ $x = \frac{1.75}{3}$

      $x = \frac{8}{9}$ $x = \frac{36}{99} = \frac{4}{11}$ $x = \frac{175}{300} = \frac{7}{12}$

Example

Round these numbers.

a 474 to 2 **significant figures**

b 45.16 to 1 **decimal place**

c 50.65 to 3 significant figures

a 474 to 2 significant figures is 470

b 45.16 to 1 decimal place is 45.2

c 50.65 to 3 significant figures is 50.7

> In example **a**, 0 has been added to 47 to make it roughly the same size as 474.
>
> In example **c**, the digit 0 is significant and it is included in the 3 significant figures.

Exercise N2

1 Give the value of the digits in bold.

   a 6.40**5**   **b** 0.2**0**4   **c** 3.3**6**27   **d** 0.000**4**6

(H p2)

2 a Convert these fractions so that they have a common denominator.

$$\frac{5}{12} \qquad \frac{4}{9} \qquad \frac{7}{18} \qquad \frac{3}{8} \qquad \frac{17}{36}$$

   b Write the fractions in order of size, smallest first.

(H p74, H+ p79)

> You need the LCM of 8, 9, 12, 18 and 36 – see N1.

3 Convert these decimals to

   a a fraction in its simplest form   **b** a percentage.

   i 0.4   **ii** 0.08   **iii** 0.18   **iv** 0.375

(H p80, 82, H+ p78)

4 Convert these percentages to

   a a fraction in its simplest form   **b** a decimal.

   i 80%   **ii** 12.5%   **iii** 42.5%   **iv** 8.75%

(H p80, 82, H+ p78)

5 Convert these fractions to

   a a decimal   **b** a percentage.

   i $\frac{9}{20}$ **ii** $\frac{19}{200}$ **iii** $\frac{23}{500}$ **iv** $\frac{37}{1000}$

(H p80, 82, H+ p78)

6 In a survey, Alan counted the number of cars in each household on his street.

The results are shown on the pie chart.

Calculate   **a** the value of x

              b the percentage for each category in the pie chart.

(H p80, H+ p78)

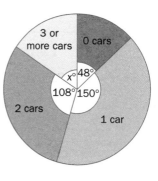

> See D2 for pie charts.

7 Fiona does a spelling test. She gets 7 words correct out of 8.

Calculate her mark as

a a fraction in its simplest form   **b** a percentage.

She now spells two more words. Her new mark is 80%.

c How many words has she now spelled incorrectly?

(H p80, H+ p78)

8 Convert these fractions to decimals.
State which fractions give recurring decimals.

   a $\frac{4}{9}$ **b** $\frac{3}{8}$ **c** $\frac{6}{11}$ **d** $\frac{5}{16}$

(H p80, H+ p76)

9 The fraction $\frac{1}{7}$ can be written as a recurring decimal.

$\frac{1}{7} = 0.142\,857\,1\ldots$

Write the next five digits in the recurring decimal.

(H p80, H+ p76)

10 Write these numbers in order of size, smallest first.

   a 63% $\frac{3}{5}$ $\frac{5}{8}$ $\frac{31}{50}$

   b 93.5% $\frac{9}{10}$ 0.905 $\frac{19}{20}$

   c 0.17 $\frac{1}{6}$ 16.5%   16%

   d 66% $\frac{2}{3}$ 0.67 $\frac{5}{8}$

   e $\frac{17}{20}$ $\frac{43}{50}$ $\frac{4}{5}$ $\frac{21}{25}$

(H p80, 82, H+ p78)

> You need to change each number to the same form, fraction, decimal or percentage – your choice.

11 Round these numbers to the required degree of accuracy.

   a 18.2 to 1 significant figure

   b 3.538 to 2 decimal places

   c 5.35 to 1 decimal place

   d 108.2 to 3 significant figures

   e 0.005 to 2 decimal places

(H p38, H+ p38)

12 Express these recurring decimals as fractions.

   a $0.\dot{5}$ **b** $0.\dot{8}\dot{1}$ **c** $0.\dot{5}0\dot{4}$ **d** $0.5\dot{6}$

(H p82, H+ p76)

13 a Change $\frac{3}{11}$ to a decimal.

   b Prove that the recurring decimal $0.\dot{3}\dot{9} = \frac{13}{33}$

(*Edexcel Ltd., 2005*) 4 marks

> Prove means you must start from
> $x = 0.393\,939\,393\ldots$
> See the example.

- You can use standard written methods to add, subtract, multiply and divide integers. Each method relies on **place value**.

- Adding a negative number is the same as subtracting a positive number. $6 + (-8) = 6 - 8$

- Subtracting a negative number is the same as adding a positive number. $6 - (-8) = 6 + 8$

- You can multiply and divide positive and negative numbers.

  $(-4) \times (-7) = +28$

Positive × positive = positive
Positive × negative = negative
Negative × positive = negative
Negative × negative = positive

- You use **BIDMAS** to decide the order of operations.

  $$(6 - 2) \times 5^2 + 1 = 4 \times 5^2 + 1$$
  $$= 4 \times 25 + 1$$
  $$= 100 + 1$$
  $$= 101$$

- You add and subtract fractions using equivalent fractions with a **common denominator**.

  $$4\frac{8}{9} + 3\frac{5}{6} = 4\frac{16}{18} + 3\frac{15}{18} = 7\frac{31}{18} = 8\frac{13}{18}$$

- To multiply fractions, you multiply the **numerators** and you multiply the **denominators**.

  $$1\frac{2}{3} \times 1\frac{3}{4} = \frac{5}{3} \times \frac{7}{4} = \frac{35}{12} = 2\frac{11}{12}$$

- To divide fractions, you use the relationship between multiplication and division.

  $$1\frac{3}{4} \div 2\frac{5}{8} = \frac{7}{4} \div \frac{21}{8} = \frac{7}{4} \times \frac{8}{21} = \frac{56}{84} = \frac{2}{3}$$

- You use **rounded numbers** to estimate the answer to a calculation

  $$\frac{475 \times 9113}{295} \approx \frac{500 \times 9000}{300}$$

Keywords
BIDMAS
Common denominator
Denominator
Numerator
Place value
Rounded numbers

See N1 for negative numbers.

Use the same rules for division.

Brackets
Powers or Indices
Division and Multiplication
Addition and Subtraction

See N2 for denominator and equivalent fractions.

See N2 for numerator.

$\div\frac{21}{8}$ is the same as $\times\frac{8}{21}$

$\dfrac{500 \times \cancel{9000}^{30}}{\cancel{300}_1} = \dfrac{15\,000}{1}$
$= 15\,000$

Example

Work out 381×462.

$$
\begin{array}{r}
381 \\
462\times \\
\hline
381 \times 400 \quad 152400 \\
381 \times 60 \quad 22860 \\
381 \times 2 \quad 762 \\
\hline
176022
\end{array}
$$

or

	300	**80**	**1**	
400	120000	32000	400	= 152400
60	18000	4800	60	= 22860
2	600	160	2	= 762
				176022

Check: $400 \times 500 = 200000$

Example

Choose three cards to complete the sum.

$\boxed{?} \times \boxed{?} \times \boxed{?} = \boxed{-6}$

$\boxed{-3}$ $\boxed{-2}$ $\boxed{-1}$ $\boxed{0}$ $\boxed{1}$

$\boxed{-3} \times \boxed{-2} \times \boxed{-1} = \boxed{-6}$

Example

Calculate the perimeter of the rectangle.
State the units of your answer.

Perimeter $= 2\frac{1}{4} + 3\frac{2}{5} + 2\frac{1}{4} + 3\frac{2}{5}$

$= 2\frac{5}{20} + 3\frac{8}{20} + 2\frac{5}{20} + 3\frac{8}{20}$

$= 10\frac{26}{20} = 11\frac{6}{20} = 11\frac{3}{10}$ cm

$2\frac{1}{4}$ cm

$3\frac{2}{5}$ cm

See S4 for perimeter.

Example

1 kilogram is about the same as $2\frac{1}{5}$ lb.

Calculate the weight in kilograms of a baby weighing 8 lb.

$8 \div 2\frac{1}{5} = \frac{8}{1} \div \frac{11}{5} = \frac{8}{1} \times \frac{5}{11} = \frac{40}{11} = 3\frac{7}{11}$ kg

Example

Ian runs for 42 minutes at $7\frac{1}{2}$ miles per hour.

Calculate the distance he ran.

See N6 for speed.

Distance = Speed × Time

$= 7\frac{1}{2} \times \frac{7}{10} = \frac{15}{2} \times \frac{7}{10} = \frac{^{3}\cancel{15}}{2} \times \frac{7}{\cancel{10}_{2}} = \frac{21}{4} = 5\frac{1}{4}$ mph

42 mins $= \frac{42}{60}$ hour

$= \frac{7}{10}$ hour

Example

Evaluate **a** $12 + 8 \times 3$ **b** $5 \times 2^3 - 3$ **c** $\sqrt{(5^2 - 4^2)}$

Use BIDMAS.

a $12 + 8 \times 3 = 12 + 24 = 36$

b $5 \times 2^3 - 3 = 5 \times 8 - 3 = 40 - 3 = 37$

c $\sqrt{5^2 - 4^2} = \sqrt{25 - 16} = \sqrt{9} = 3$

Exercise N3

1 Calculate

  a 123×57 **b** $799 \div 17$ **c** $7847 \div 59$

  (H p46)

2 The rows, columns and diagonals of a magic square add to the same number.

  No number can be repeated.

  Copy and complete the magic square.

  (H p4)

2		0
	−1	
		−4

3 You are given six cards.

  a Use two cards to complete the sum to give the greatest possible answer.

    $\boxed{} \times \boxed{} = \boxed{\text{Greatest possible answer}}$

  b From all six cards, select two cards with the least possible product.

  (H p6)

2	1	0
−1	−2	−3

A product is the result of multiplication.

4 Sophie and Rosie work out this sum: $\dfrac{16 + 14}{10 + 5}$

  Sophie's answer is 2.

  Rosie's answer is 8.

  Who is correct? Give a reason for your choice.

  (H p206, H+ p212)

5 Evaluate

  a $16 + 4 \times 5$ **b** $(6 - 8)^3$ **c** $5^2 - \dfrac{20}{4}$

  (H p206, H+ p212)

Evaluate means work out.

6 Estimate an approximate answer to each sum.

  a $\dfrac{4920}{19}$ **b** $\dfrac{391 \times 88}{31 \times 48}$ **c** $\dfrac{27}{\frac{1}{4}}$

  (H p38)

First round the numbers to one significant figure – see N2.

7 Calculate

  a $2\frac{5}{12} + 1\frac{5}{6}$ **b** $4\frac{7}{8} - 2\frac{3}{10}$ **c** $2\frac{2}{3} - \frac{4}{5}$

  (H p76, H+ p74)

8 Calculate the volume of the cuboid.

  State the units of your answer.

  (H p78, H+ p74)

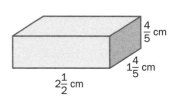

$\frac{4}{5}$ cm

$1\frac{4}{5}$ cm

$2\frac{1}{2}$ cm

9 Calculate these and simplify if possible.

For fractions simplify means cancel.

a $18 \times \frac{4}{5}$ **b** $\frac{3}{4} \times 1\frac{2}{3}$ **c** $2\frac{1}{2} \div \frac{15}{16}$

d $3\frac{1}{2} \div 2\frac{4}{5}$ **e** $\sqrt{1\frac{7}{9}}$ **f** $36 \div \frac{4}{7}$

(H p78, H+ p74)

10 The size of a drill bit is half way between $\frac{1}{2}$ and $\frac{9}{16}$ inches.

Calculate the size of this drill bit.

(H p76, 78, H+ p74)

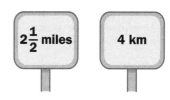

11 $2\frac{1}{2}$ miles is about the same as 4 kilometres.

Calculate the number of kilometres in one mile.

(H p78, 152, 286, H+ p74)

12 A car journey of 230 km lasts 1 hour and 55 minutes.

a Calculate the speed in km per hour.

b The speed limit is 110 kmph.

Does the car break the speed limit?

(H p78, 152, 286, H+ p74)

13 a Work out the value of $\frac{2}{3} \times \frac{3}{4}$.

Give your answer as a fraction in its simplest form.

b Work out the value of $1\frac{2}{3} \times 2\frac{3}{4}$.

Give your answer as a fraction in its simplest form.

(Edexcel Ltd., 2005) 5 marks

14

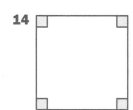

Diagrams NOT drawn to scale.

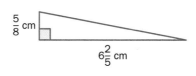

The area of the square is 18 times the area of the triangle.

Work out the **perimeter** of the square.

See S4 for perimeter.

(Edexcel Ltd., 2004) 5 marks

Keywords
Appropriate degree of accuracy
Decimal places
Significant figures
Trial and improvement
Upper and lower bounds

- You need to interpret the calculator display when you are doing money calculations. $\boxed{4\;15.3}$ means £415.30

- You can round numbers to their approximate size using
  - **decimal places** 40.835 rounded to 2 decimal places is 40.84
  - **significant figures** 7.04 rounded to 2 significant figures is 7.0

  Write 2 digits.

- Some answers must be given to an **appropriate degree of accuracy.**

  £2.673 456 is interpreted as £2.67.

  4.167 485 9 cm might be interpreted as 4.2 cm.

- In trigonometry, calculated angles are usually given to at least one decimal place.

  See S7 for trigonometry.

- The accuracy of the numbers in the problem will determine how accurately you should give your answers.

  Given the lengths 4.1 cm and 8.5 cm in a right-angled triangle, the hypotenuse is 9.4 cm, not 9.437 160 6 cm.

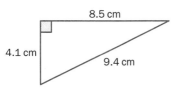

- You can estimate the answer to a calculation by using rounded numbers.

  $$\frac{2.7 + 780}{49.8 \times 0.85} \approx \frac{800}{50 \times 1} = \frac{800}{50} = 16$$

  Ignore the relatively small 2.7, when adding to 780.

- You multiply decimals using the rules of multiplication and division by powers of 10.

  $8.5 \times 3.7 = 85 \times 37 \div 10 \div 10 = 3145 \div 100 = 31.45$

  Always check your answer by approximating.

  $9 \times 4 = 36$

  See N8 for × and ÷ by powers of 10.

  First remove the decimal points and calculate. Then insert the decimal point in the appropriate position.

- You divide by a decimal using the rules of multiplication by powers of 10.

  You must divide by an integer.

  $45.22 \div 1.7$ is the same as $452.2 \div 17$

  Always check your answer by approximating.

  $45 \div 2 = 22.5$

  See N1 for integers.
  See N8 for × by powers of 10.

  Multiply both numbers by 10 or 100 until you can divide by an integer.

- You can find an approximate solution to some problems using **trial and improvement** techniques.

  $\sqrt{85} = 9.2$ (to 1 dp)

  See the example and A3.

- The **upper and lower bounds** are the two limits that a number can take.

  If 18 cm is a measurement rounded to the nearest centimetre, the upper bound is 18.5 cm and the lower bound is 17.5 cm.

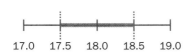

Example

Calculate the square root of 85 by trial and improvement.

Give your answer to 1 decimal place.

See N5 for square roots.

Number	(Number)^2	
8	64	too small
9	81	too small
10	100	too big
9.5	90.25	too big
9.2	84.64	too small
9.3	86.49	too big
9.25	85.5625	too big

$\sqrt{85}$ = 9.2 (to 1 dp)

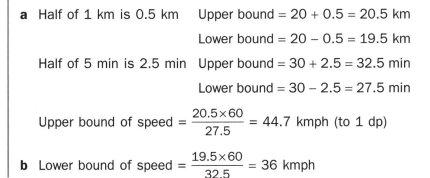

too small   too big   too big

9.2   9.25   9.3

Example

A machine can make 600 toys in one hour.

Calculate the time taken to make 1040 toys.

Time taken = $\frac{1040}{600}$ = 1.7$\dot{3}$ hours = 1.7$\dot{3}$ × 60 minutes

= 104 min or 1 hour 44 min

Example

Angela cycles 20 km in 30 minutes.

The distance is correct to the nearest km.

The time is correct to the nearest 5 minutes.

Calculate   **a** the upper bound of Angela's speed in km per hour

b the lower bound of Angela's speed in km per hour.

Give each answer to an appropriate degree of accuracy.

a Half of 1 km is 0.5 km   Upper bound = 20 + 0.5 = 20.5 km

Lower bound = 20 − 0.5 = 19.5 km

Half of 5 min is 2.5 min   Upper bound = 30 + 2.5 = 32.5 min

Lower bound = 30 − 2.5 = 27.5 min

Upper bound of speed = $\frac{20.5 \times 60}{27.5}$ = 44.7 kmph (to 1 dp)

b Lower bound of speed = $\frac{19.5 \times 60}{32.5}$ = 36 kmph

See N6 for speed.

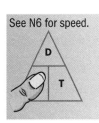

Exercise N4

1 Using trial and improvement, calculate $\sqrt[3]{36}$ correct to 1 decimal place.

See N5 for cube roots.

(H p222)

2 Ben has been asked to calculate $19.6 \div 0.07$.

    a Copy and complete this sentence.

        $19.6 \div 0.07$ is the same as .... ÷ 7

    b What answer should Ben get?

(H p46)

3 The rectangular base of a garage measures 4.5 m by 2.3 m.

    a Calculate the area of the base.

        State the units of your answer.

    It costs £1.50 to paint each square metre of surface.

    b Calculate the cost of painting the garage base.

(H p46)

Do not round your area answer. Only round when you have found the cost of the paint.

4 Round these numbers to the required degree of accuracy.

    a 48.35 to 1 decimal place

    b 304 to 2 significant figures

    c 20.688 to 3 significant figures

    d 0.909 to 2 decimal places

    e 0.0097 to 3 decimal places

(H p38, H+ p38)

5 a Calculate $\frac{4}{7}$ of 3.8.

    b Give your answer to 1 decimal place.

(H p38, H+ p38)

6 a By rounding each number to 1 significant figure, estimate

        $\dfrac{3.85 + 9.42}{2.31 \times 1.56}$

See N3 for BIDMAS.

    b Use your calculator to work out the exact value of the sum.

        Write all the figures on your calculator display.

    c Write your answer to an appropriate degree of accuracy.

(H p38, 206, H+ p206)

7 Convert these journey times into minutes

    a 1.25 hours   **b** 2.15 hours   **c** $3.\dot{6}$ hours

8 Calculate the least and greatest number of people that could have arrived at each airport. Each number has been rounded correct
to 2 significant figures.

a Heathrow   14 000

b Stansted   8500

c Gatwick   13 000

d Luton   7400

e London City   1900

(H p214, H+ p40)

9 Find the **upper** and **lower bounds** of the **rounded** numbers.

a 490 km has been rounded to the nearest 10 km

b 16 seconds has been rounded to the nearest second

c 3 m has been rounded to the nearest cm

d 8.5 cm has been rounded to the nearest 5 mm

e 25.0 seconds has been rounded to the nearest tenth of a second

(H p214, H+ p40)

10 A piece of wood has a mass of 250 g, to the nearest 10 grams.

The volume of the wood is 24 cm^3, to the nearest cm^3.

Calculate the greatest and least possible values for the density of the wood.

Give your answers to a suitable degree of accuracy.

(H p214, H+ p40)

See N6 for density.

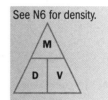

11 Each side of a regular pentagon has a length of 101 mm, correct to the nearest millimetre.

a Write the **least** possible length of each side.

b Write the **greatest** possible length of each side.

(*Edexcel Ltd., 2004*) 2 marks

12

MAXIMUM LOAD
1200 kg

Weight 60 kg

Peter transports metal bars in his van.

The van has a safety notice 'Maximum Load 1200 kg'

Each metal bar has a label 'Weight 60 kg'.

For safety reasons Peter assumes that

1200 is rounded correct to 2 significant figures

and 60 is rounded correct to 1 significant figure.

Calculate the greatest number of bars that Peter can **safely** put into the van if his assumptions are correct.

(*Edexcel Ltd., 2005*) 4 marks

- The **power** of a number tells you how many times the number must be multiplied by itself.

  $8^4 = 8 \times 8 \times 8 \times 8 = 4096$ 4 is the power.

- A square number is the result of multiplying an integer by itself. 81 is a square number, as $9 \times 9 = 81$.

- A cube number is the result of multiplying an integer by itself and then by itself again. 343 is a cube number, as $7 \times 7 \times 7 = 343$.

- The **square root** of a number multiplied by itself makes the number. 6 is the square root of 36, as $6 \times 6 = 36$.

- A positive number has both a positive square root and a negative square root. The square roots of 64 are 8 and −8.

- The **cube root** of a number multiplied by itself and then by itself again makes the number. 9 is the cube root of 729, as $9 \times 9 \times 9 = 729$.

- A fractional power represents a root.

  $9^{\frac{1}{2}} = \sqrt{9} = 3$ and $125^{\frac{1}{3}} = \sqrt[3]{125} = 5$

- Any number raised to the power of zero is equal to 1.

- The **reciprocal** of a number is 1 divided by that number.

  The reciprocal of 2 is $\frac{1}{2}$.

- A negative power represents the reciprocal of the number.

  $5^{-1} = \frac{1}{5}$ and $5^{-2} = \frac{1}{5^2} = \frac{1}{25}$

- When you multiply powers of the same number, you add the powers.

  $7^4 \times 7^2 = 7^{4+2} = 7^6$

- When you divide powers of the same number, you subtract the powers.

  $4^3 \div 4^2 = 4^{3-2} = 4^1$

- A **surd** is the square root of an integer, which produces an irrational number.

  $\sqrt{2}$ and $\sqrt{7}$ are examples of surds.

- You can simplify some expressions using factors.

  $\sqrt{28} = \sqrt{4 \times 7} = \sqrt{4} \times \sqrt{7} = 2 \times \sqrt{7} = 2\sqrt{7}$
  $\sqrt{3} \times \sqrt{3} = \sqrt{3 \times 3} = \sqrt{9} = 3$

- You can simplify some fractions by multiplying by '1'.

  $\frac{1}{\sqrt{3}} = \frac{1}{\sqrt{3}} \times \left(\frac{\sqrt{3}}{\sqrt{3}}\right) = \frac{1 \times \sqrt{3}}{\sqrt{3} \times \sqrt{3}} = \frac{\sqrt{3}}{\sqrt{9}} = \frac{\sqrt{3}}{3}$

Keywords
Cube root
Power
Reciprocal
Square root
Surd

$9^2 = 81$

$7^3 = 343$

$\sqrt{}$ means square root. $\sqrt{36} = 9$

$8 \times 8 = 64$
$-8 \times -8 = 64$

$\sqrt[3]{}$ means cube root. $\sqrt[3]{729} = 9$

$64^{\frac{2}{3}} = \left(64^{\frac{1}{3}}\right)^2 = 4^2 = 16$

$4^0 = 1$

$2 \times \frac{1}{2} = 1$

$4^1 = 4$

See N2 for irrational numbers.

$\left(\sqrt{3}\right)^2 = 3$

Multiplying by 1 does not alter the value of the expression.
$\left(\frac{\sqrt{3}}{\sqrt{3}}\right) = 1$

Example

Calculate

a 7^5 **b** 8^{-1} **c** 5^0 **d** 6^{-3}

a $7^5 = 7 \times 7 \times 7 \times 7 \times 7 = 16\,807$

b $8^{-1} = \frac{1}{8}$

c $5^0 = 1$

d $6^{-3} = \frac{1}{6^3} = \frac{1}{216}$

Example

Write each of these in the form 4^n.

a 64   **b** $\frac{1}{4}$ **c** $\frac{1}{16}$ **d** 2

a $64 = 4 \times 4 \times 4 = 4^3$ **b** $\frac{1}{4} = 4^{-1}$

c $\frac{1}{16} = \frac{1}{4} \times \frac{1}{4} = 4^{-1} \times 4^{-1} = 4^{-2}$ **d** $2 = \sqrt{4} = 4^{\frac{1}{2}}$

Example

Calculate

a $9^{\frac{3}{2}}$ **b** $1000^{\frac{2}{3}}$ **c** $36^{-\frac{1}{2}}$ **d** $\left(\frac{16}{25}\right)^{\frac{1}{2}}$

a $9^{\frac{3}{2}} = \left(9^{\frac{1}{2}}\right)^3 = 3^3 = 27$ **b** $1000^{\frac{2}{3}} = \left(1000^{\frac{1}{3}}\right)^2 = 10^2 = 100$

c $36^{-\frac{1}{2}} = \frac{1}{36^{\frac{1}{2}}} = \frac{1}{6}$ **d** $\left(\frac{16}{25}\right)^{\frac{1}{2}} = \frac{4}{5}$

Example

Simplify these expressions.

a $\sqrt{2} \times \sqrt{5}$ **b** $\frac{14}{\sqrt{7}}$ **c** $5\sqrt{2} - 3\sqrt{2}$ **d** $\left(\sqrt{5}\right)^3$

a $\sqrt{2} \times \sqrt{5} = \sqrt{10}$

b $\frac{14}{\sqrt{7}} = \frac{14}{\sqrt{7}} \times \left(\frac{\sqrt{7}}{\sqrt{7}}\right) = \frac{14\sqrt{7}}{\sqrt{49}} = \frac{14\sqrt{7}}{\sqrt{7}} = 2\sqrt{7}$

c $5\sqrt{2} - 3\sqrt{2} = 2\sqrt{2}$

d $\left(\sqrt{5}\right)^3 = \sqrt{5} \times \sqrt{5} \times \sqrt{5} = \sqrt{25} \times \sqrt{5} = 5 \times \sqrt{5} = 5\sqrt{5}$

Example

Write these expressions in the form $a\sqrt{b}$.

a $\sqrt{32}$ **b** $\sqrt{75}$ **c** $\sqrt{500}$ **d** $\frac{\sqrt{12}}{\sqrt{3}}$

a $\sqrt{32} = \sqrt{16 \times 2} = \sqrt{16} \times \sqrt{2} = 4 \times \sqrt{2} = 4\sqrt{2}$

b $\sqrt{75} = \sqrt{25 \times 3} = \sqrt{25} \times \sqrt{3} = 5 \times \sqrt{3} = 5\sqrt{3}$

c $\sqrt{500} = \sqrt{100 \times 5} = \sqrt{100} \times \sqrt{5} = 10 \times \sqrt{5} = 10\sqrt{5}$

d $\frac{\sqrt{12}}{\sqrt{3}} = \frac{\sqrt{4 \times 3}}{\sqrt{3}} = \frac{\sqrt{4} \times \sqrt{3}}{\sqrt{3}} = \frac{2 \times \sqrt{3}}{\sqrt{3}} = 2\sqrt{\left(\frac{3}{3}\right)} = 2\sqrt{1} = 2$

Exercise N5

1 **a** Calculate the value of each card.

   b Put the cards in order, smallest first.

   (H p170)

4^5 3^6 2^9 5^4

2 Write each of these in the form 8^n.

   a $8^5 \times 8^3$ **b** $8^9 \div 8^2$ **c** $\dfrac{8^{10}}{8^2 \times 8^5}$ **d** $\dfrac{1}{8^2}$

   (H p172, H+ p170)

3 Rewrite these numbers as fractions without indices.

   a 5^{-1} **b** 10^{-2} **c** 2^{-4} **d** 5^{-3}

   (H p174, H+ p172)

4 Find the value of n in each equation.

   a $8^n = 512$ **b** $10^n = \dfrac{1}{10}$ **c** $6^n = 1$ **d** $144^n = 12$

   e $n^{\frac{1}{3}} = 3$ **f** $n^{-2} = \dfrac{1}{25}$ **g** $n^{-\frac{1}{4}} = \dfrac{1}{3}$ **h** $n^{\frac{3}{4}} = 27$

   (H p174, H+ p172)

5 Calculate

   a $36^{\frac{1}{2}}$ **b** $256^{\frac{1}{8}}$ **c** $8^{\frac{2}{3}}$ **d** $32^{\frac{4}{5}}$

   (H+ p172, H+ p172)

6 Simplify

   a $\sqrt{72}$ **b** $\sqrt{63}$ **c** $\sqrt{180}$ **d** $\sqrt{192}$

   (H p208, H+ p176)

7 Simplify

   a $\sqrt{20} - \sqrt{5}$ **b** $\sqrt{98} - 3\sqrt{2}$ **c** $10\sqrt{2} - 2\sqrt{8}$ **d** $5\sqrt{3} - \sqrt{48}$

   (H p208, H+ p176)

8 Simplify

   a $\dfrac{3\sqrt{8}}{\sqrt{2}}$ **b** $\dfrac{5\sqrt{3}}{\sqrt{12}}$ **c** $\dfrac{\sqrt{50}}{2\sqrt{10}}$ **d** $\dfrac{6}{\sqrt{3}}$

   (H+ p176, p208)

9 Simplify these.

   a $(8 + \sqrt{5})(8 - \sqrt{5})$ **b** $(\sqrt{2} + 1)^2$

   c $\dfrac{(3 + \sqrt{2})(3 - \sqrt{2})}{\sqrt{7}}$ **d** $\dfrac{1}{1 + \sqrt{5}}$

   (H p208, H+ p176)

10 A cuboid has dimensions $\sqrt{2}$ cm, $\sqrt{2}$ cm and $\left(\sqrt{3} + \sqrt{12}\right)$ cm.

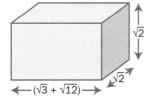

   a Calculate the volume of the cuboid.

     Fully simplify your answer.

   A cube of side $\sqrt{3}$ cm is cut out of the cuboid.

   b Calculate the volume of the remaining solid.

     (H p208, H+ p176)

11 a Write as a power of 5

    i $5^4 \times 5^2$

    ii $5^9 \div 5^6$

   b $2^x \times 2^y = 2^{10}$ and $2^x \div 2^y = 2^4$
     Work out the value of x and the value of y.

    (Edexcel Ltd., 2005) 5 marks

> See A3 for simultaneous equations.

12 Work out

   a 4^0

   b 4^{-2}

   c $16^{\frac{2}{3}}$

  (Edexcel Ltd., 2003) 3 marks

13 Work out $\dfrac{\left(5 + \sqrt{3}\right)\left(5 - \sqrt{3}\right)}{\sqrt{22}}$

Give your answer in its simplest form.

(Edexcel Ltd., 2003) 3 marks

14 a Find the value of $16^{\frac{1}{2}}$.

   b Given that $\sqrt{40} = k\sqrt{10}$, find the value of k.

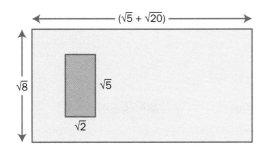

A large rectangular piece of card is $\left(\sqrt{5} + \sqrt{20}\right)$ cm wide and $\sqrt{8}$ cm long.

A small rectangle $\sqrt{2}$ cm wide and $\sqrt{5}$ cm long is cut out of the piece of card.

   c Express the area of the card that is left as a percentage of the area of the large rectangle.

(Edexcel Ltd., 2004) 6 marks

Keywords
Direct proportion
Rate
Ratio
Scale
Unitary method

● You can use a **ratio** to compare the size of two (or more) quantities.

The ratio 1:4 means the second quantity is 4 times larger than the first.

● Some ratios can be simplified by cancelling.

36:45 is equivalent to 4:5.

● A **scale** is a ratio expressed in the form 1:*n*.

A map scale could be 1:50 000.

> Real-life is 50 000 times larger than the map.
>
> See S6 for enlargements.

● You can divide a quantity in a given ratio.

180° divided in the ratio 2:3:4 is 40°, 60° and 80°.

> Check
> 40° + 60° + 80° = 180°

● Two quantities are in **direct proportion**, when the ratio between the quantities is constant.

As one quantity increases, the other quantity increases in the same proportion.

As one quantity decreases, the other quantity decreases in the same proportion.

> See A4 for inverse proportion.

In the graph, the cost (*C*) is directly proportional to the number of sweets (*n*) bought.

$C \propto n$ or $C = kn$

where *k* is a constant.

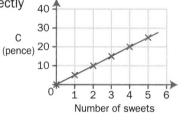

> $C = kn$
> See A6 for
> $y = mx + c$

● You can use the **unitary method** to solve direct proportion problems.

This method first finds the value of 1 unit.

● A **rate** compares a quantity with one unit of another quantity.

Speed can be measured in kilometres per hour.

Density can be measured in kilograms per cubic metre.

> See A8 for speed.

● You can use conversion rates and exchange rates to convert between two different quantities.

1 kilogram (kg) = 2.2 pounds (lb)

10 kilograms (kg) = 22 pounds (lb)

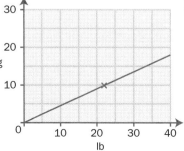

> See A8 for conversions with graphs.

Example

Mortar is made from cement and sand in the ratio $1:4$.

What percentage of mortar consists of sand?

Mortar = 1 + 4 = 5 parts

Sand = 4 parts

cement	sand	sand	sand	sand

Proportion of sand = $\frac{4}{5}$ = 80%

Example

Safina and Violet work at a restaurant.
They share the tips according to the number of hours each works.
Safina works 3 hours a day. Violet works 5 hours a day.
Safina's share works out to be £34.50.

a How much should Violet expect?

b Calculate the total amount of tips.

a Safina has 3 parts.

$£34.50 \div 3 = £11.50$

Violet has 5 parts.

$£11.50 \times 5 = £57.50$

b $£34.50 + £57.50 = £92$

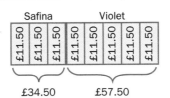

Check
£92 ÷ 8 = £11.50

Example

A cube of side 2 cm weighs 63 g.

a Calculate the density of the material used to make the cube in g per cm^3.

b Which metal do you think has been used to make the cube?

Material	Density (g/cm³)
Copper	8.9
Lead	11.4
Aluminium	2.6
Brass	8.6
Iron	7.9

a Volume = $2 \times 2 \times 2 = 8$ cm³

Density = $63 \div 8 = 7.875$ g/cm³

b Iron is the most likely.

Example

A map scale is $1:25\,000$.

a A road measures 8 cm on the map.

How long is the road in real-life?

b Julie wants to walk 10 miles.

Calculate the distance, in centimetres,
on the map that represents 10 miles in real-life.

Scale 1:25 000

a $8 \times 25\,000$ cm = $200\,000$ cm = 2000 m = 2 km

b 10 miles = 16 kilometres

$= 16\,000$ metres

$= 1\,600\,000$ cm

$\frac{1\,600\,000}{25\,000} = 64$ cm

5 miles ≡ 8 kilometres

Exercise N6

1 A map has a scale of 4 cm represents 2 km.

   a Rewrite the scale as a ratio in its simplest form.

   b Calculate the distance between villages if they are 3.2 cm apart on the map.

   c If the length of a straight road is 1.5 km, how long will the road appear on the map?

   (H+ p154)

2 Kayla goes on holiday to France.

   She buys a DVD player for €25 using her credit card.

   Her credit card statement says she has spent £16.78.

   Calculate the exchange rate between pounds (£) and euros (€) by finding the value of £1 in euros.

   Give your answer to a suitable degree of accuracy.

   (H p150)

3 A car travels 24 km in 40 minutes.

   a Calculate the average speed in kilometres per hour.

   b If the car maintains this average speed, how long will it take to travel 90 km?

   (H p152, 286)

4 In 1954, Roger Bannister became the first person to run a mile in under 4 minutes.

   His time was 239.4 seconds.

   a Calculate his speed in metres per second.

   b Calculate the time he would have taken to run 1500 metres.

   (H p152, 286)

5 miles ≡ 8 kilometres

5 A wheelbarrow holds 70 litres.

   It is filled with earth that has a density of 0.6 kg/litre.

   Calculate the weight of the earth in the wheelbarrow.

   State the units of your answer.

   (H p152, 286)

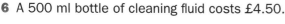

6 A 500 ml bottle of cleaning fluid costs £4.50.

   How much should 700 ml of the same fluid cost?

   (H p148, H+ p148)

7 The prices and sizes of three jars of coffee are shown.

Which jar of coffee is the best value?

You must show your working.

(H p148, H+ p148)

8 Two friends, Emma and Ruth, decide to buy a car between them.
The car costs £2400.
Emma pays £1000 and Ruth pays the rest.

a Work out the ratio of Emma's payment to Ruth's payment.
Give your answer in its simplest form.

Two years later, they sell the car for £960.

b How much should Emma and Ruth each get?

(H p304, H+ p154)

9 Jim, William and Harry go to the Bingo.

They share the winnings in the ratio 5:3:2 as Jim buys 5 cards,
William buys 3 cards and Harry buys 2 cards.

Harry receives £64.

a Work out the total winnings.

b Calculate the proportion of each person's winnings as a
percentage of the total winnings.

(H p304, 308, H+ p154)

4			30		52	61	70	81
	10	21		43				85
9		37	48		68		86

See N7 for percentage change.

10 1 foot is approximately 0.3048 metres.

a Calculate the value of 3 feet in centimetres.
Give your answer to a suitable degree of accuracy.

b Find the number of feet in one metre, giving your answer to
3 decimal places.

c Which is larger, 3 feet or 1 metre?

(H+ p150)

11 Three women earned a total of £36.
They shared the £36 in the ratio 7 : 3 : 2.
Donna received the largest amount.

a Work out the amount Donna received.

A year ago, Donna weighed 51.5 kg.
Donna now weighs 8.5% less.

b Work out how much Donna now weighs.
Give your answer to an appropriate degree of accuracy.

(*Edexcel Ltd., 2005*) 7 marks

See N7 for percentage calculations.

- You calculate a **percentage** of an amount using

    equivalent decimals

    35% of 40 kg = 0.35 × 40 kg

  or   mental arithmetic, for example, using 10% to find other percentages

    17.5% = 10% + 5% + 2.5%

- You increase an amount by a percentage by

    working out the increase and adding it to the original amount

    Increase 60 km by 5%, 60 km + 3 km = 63 km

  or   using a **multiplier**

    Increase 60 km by 5%, 60 km × 1.05 = 63 km

- You decrease an amount by a percentage by

    working out the decrease and subtracting it from the original amount

    Decrease £30 by 15%, £30 − £4.50 = £25.50

  or   using a multiplier

    Decrease £30 by 15%, £30 × 0.85 = £25.50

- To calculate **simple interest** you multiply the interest earned at the end of the year by the number of years.

    The interest is the same each year and remains unaltered.

    4% of £200 = £8

- To calculate **compound interest** you work out the interest earned at the end of each year.

    The interest is not the same each year and increases each year.

    4% of £200 = £8

    4% of £208 = £8.32

    Using a multiplier,

    After 1 year   £200 × 1.04 = £208

    After 2 years   £208 × 1.04 = £216.32

- You can use a multiplier to describe exponential growth and decay.

    The value of the above investment is £200 × $(1.04)^n$ where n is the number of years.

- You can work out the **percentage change** by calculating the change as a percentage of the original value.

    Examples of percentage change are percentage loss, percentage profit and rate of inflation.

- In a **reverse percentage** problem, you are given an amount **after** a percentage change, and you have to find the original amount.

    'A construction job costs £1410 **after** VAT has been added.

    Work out the price of the job **before** VAT was added.'

Keywords

Compound interest
Multiplier
Percentage
Percentage change
Reverse percentage
Simple interest

See N2.
$10\% = \dfrac{1}{10}$

5% of 60 km = 3 km

See N2.
5% = 0.05

See N2.
15% = 0.15

Year 1   interest = £8
Year 2   interest = £8
and so on.

Year 1   interest = £8
Year 2   interest = £8.32
and so on.

£200 × $(1.04)^2$ = £216.32

index or power

You can also use a multiplier to find the percentage change.

See the last example.

Example

A shop gives a discount of 10% every Wednesday.

During a special promotional week, a further 5% is taken off.

Roy says that this means 15% is taken off the normal price.

He is WRONG. Explain why.

Choose an amount of money, say £100

Discounted price of 10% on £100 $0.9 \times £100$ = £90

Discounted price of 5% on £90 $0.95 \times £90$ = £85.50 ⎫
                                                                  ⎬ Not the same
Discounted price of 15% on £100 $0.85 \times £100$ = £85 ⎭

Example

The 2005/2006 Champions of the Football Premiership were Chelsea.

They played 38 matches, winning 29 of them.

Calculate the percentage of matches won.

See N2 for equivalents.

$\frac{29}{38}$ = 0.763 = 76.3% (to 1 decimal place)

Example

The number of wasps in a nest increases by 20% **each day**.

Initially there are 500 wasps.

a Calculate the number of wasps after **i** 1 day

                                          ii 4 days

b Find the number that you can multiply by 500 to give the number of wasps after 1 week.

---

a i $500 \times 1.2 = 600$ wasps

ii $500 \times 1.2 \times 1.2 \times 1.2 \times 1.2 = 500 \times (1.2)^4$
                                    $= 1037$ (to nearest wasp)

b $(1.2)^7 = 3.5831808$

On your calculator enter
$1.2 \; x^y \; 7 =$

Example

VAT (Value Added Tax) is charged at 17.5%.

A construction job costs £1410 **after** VAT has been added.

Work out the price **before** VAT was added.

Reverse percentage question

17.5% = 0.175

$? \times 1.175 = 1410$

$? = 1410 \div 1.175 = £1200$

Check
$0.175 \times £1200 = £210$
$£1200 + £210 = £1410$

Exercise N7

1 Without using a calculator, work out the VAT (17.5%) payable on a bill for £30.

(H p256, H+ p80)

2 An image on a computer measures

   2592 pixels by 1944 pixels

The image is increased by 12%.

Calculate the dimensions of the new image.

Give your answers to a suitable degree of accuracy.

(H p258, H+ p80)

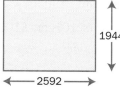
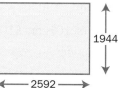

3 Lake Volta in Ghana has a surface area of 8.4×10^9 m².

In drought conditions, the surface area reduces by 5%.

Calculate the surface area after a 5% reduction.

Give your answer in standard form.

(H p258, H+ p80)

See N8 for standard form.

4 Alan is a brick-layer and earns £360 per week.

In one week, he expects to lay 1500 bricks.

His boss decides to give Alan a pay rise of £12, but only if he can increase the number of bricks he can lay by the same percentage.

Calculate   a   the percentage increase of his wage

       b   the number of extra bricks he must lay each week.

(H p262, H+ p80)

5 Brendon works at a market.

He buys a lawn mower for £35 and then sells it for £50.

What is his percentage profit?

(H p262, H+ p80)

6 Following a 7% price reduction, a printer costs £46.50.

What was the price of the printer **before** the reduction?

(H p310, H+ p82)

Reverse percentage question

7 Sylvia invests £1000 at her bank.

Each year, interest is paid at 6%.

a Calculate the amount of money in her account after 3 years.

b Find the number of years before her investment reaches £1500.

(H p260, H+ p152)

Compound interest

8 Rosa's car is worth £600.

The car loses 15% of its value each year.

The value, £V, of the car after t years is given by the equation

$V = 600 \times (q)^t$ where q is a constant.

a State the value of q.

b Calculate the value of the car after 2 years.

c After how many years will the value of the car be below £300?

(H+ p152)

9 In a sale, normal prices are reduced by 20%.

Andrew bought a saddle for his horse in the sale.
The sale price of the saddle was £220.

Calculate the normal price of the saddle.

(*Edexcel Ltd., 2005*) 3 marks

SALE

20% OFF

10 A company bought a van that had a value of £12 000.

Each year the value of the van depreciates by 25%.

a Work out the value of the van at the end of three years.

The company bought a new truck.
Each year the value of the truck depreciates by 20%.
The value of the new truck can be multiplied by a single number to find its value at the end of four years.

b Find this single number as a decimal.

(*Edexcel Ltd., 2004*) 5 marks

11 Bill invests £500 on 1 January 2004 at a compound interest rate of R% per annum.

The value, £V, of this investment after n years is given by the formula

$V = 500 \times (1.045)^n$

a Write the value of R.

b Use your calculator to find the value of Bill's investment after 20 years.

(*Edexcel Ltd., 2005*) 3 marks

- You multiply or divide decimal numbers by 10, 100, 1000, ... by moving each digit 1, 2, 3, ... places to the left or right.

$$87.2 \times 100 = 8720$$

$$695.8 \div 10 = 69.58$$

- You can use **powers of 10** to write large and small numbers in **standard form**.

4.8×10^3 is written in standard form.

- In standard form, you write a number as $A \times 10^n$

  – A is between 1 and 10 (but not including 10)

  – n is an integer.

- Small numbers are rewritten using **negative** powers of 10.

7×10^{-3} is written in standard form.

- You may need to enter a standard form number in your calculator.

3.56×10^8 is entered as

The calculator display should read

or

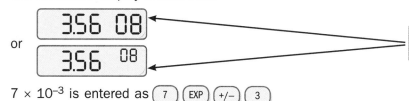

7×10^{-3} is entered as

- On some calculators you can change a number to standard form using ⟨ = ⟩

⟨ 0.00075 ⟩ ⟨ = ⟩ gives 7.5×10^{-4}

- You can multiply standard form numbers using the rules of powers.

$$(8.2 \times 10^2) \times (5 \times 10^3) = 41 \times 10^5 = 4.1 \times 10^6$$

- You can divide standard form numbers using the rules of powers.

$$(8.2 \times 10^6) \div (5 \times 10^2) = 1.64 \times 10^4$$

Keywords
Negative
Powers of 10
Standard form

Multiplying by 100 gives a larger answer.
Dividing by 10 will give a smaller answer.

See N5 for powers.

$4.8 \times 10 \times 10 \times 10$

$1 \leqslant A < 10$

See N1 for integers.

$7 \div 10 \div 10 \div 10$

On some calculators EXP is written EE.

Never write this. You will lose marks.
Always write 3.56×10^8

$7.5 \div 10 \div 10 \div 10 \div 10$

Multiply the numbers, but add the powers.

Divide the numbers, but subtract the powers.

Example

Amy decided to work out $4.7 \div 4.7 \div 4.7$.

This is the display on Amy's calculator. $4.7\text{-}03$

This is wrong. Explain why.

$4.7\text{-}03$ means $4.7 \times 10^{-3} = 4.7 \div 10 \div 10 \div 10 = 0.0047$

Example

If $p = 4.25 \times 10^2$ and $q = 3.7 \times 10^{-1}$

calculate   **a** pq **b** $\frac{p}{q}$ **c** $p + q$

Give your answers in standard form to 2 significant figures.

a $pq = 4.25 \times 10^2 \times 3.7 \times 10^{-1}$

$\quad = (4.25 \times 3.7) \times (10^2 \times 10^{-1})$

$\quad = 15.725 \times 10^1$

$\quad = 1.5725 \times 10^2$

$\quad = 1.6 \times 10^2$ (to 2 significant figures)

b $\frac{p}{q} = (4.25 \times 10^2) \div (3.7 \times 10^{-1})$

$\quad = (4.25 \div 3.7) \times (10^2 \div 10^{-1})$

$\quad = 1.149 \times 10^3$

$\quad = 1.1 \times 10^3$ (to 2 significant figures)

c $p + q = (4.25 \times 10^2) + (3.7 \times 10^{-1})$

$\quad = 425 + 0.37$

$\quad = 425.37$

$\quad = 4.2537 \times 10^2$

$\quad = 4.3 \times 10^2$ (to 2 significant figures)

Example

Using standard form, estimate the value of

a 2143×0.0006 **b** $9045 \div 0.054$ **c** $\frac{71 \times 0.04}{833 \times 471}$

a $2143 \times 0.0006 \approx 2000 \times 0.0006 = 2 \times 10^3 \times 6 \times 10^{-4}$
$\quad\quad = 12 \times 10^{-1} = 1.2$

b $9045 \div 0.054 \approx 9000 \div 0.05 = (9 \times 10^3) \div (5 \times 10^{-2})$
$\quad\quad = 1.8 \times 10^5 = 180\,000$

c $\frac{71 \times 0.04}{833 \times 471} \approx \frac{70 \times 0.04}{800 \times 500} = (7 \times 10^1 \times 4 \times 10^{-2}) \div (8 \times 10^2 \times 5 \times 10^2)$
$\quad\quad = (28 \times 10^{-1}) \div (40 \times 10^4)$
$\quad\quad = 0.7 \times 10^{-5}$
$\quad\quad = 0.000\,007$

Exercise N8

1 Copy and complete this table. The first line is done for you.

Power of 10	Meaning	Number
10^4	$10 \times 10 \times 10 \times 10$	10 000
10^3		
10^2		
10^1		
10^0		
10^{-1}		
10^{-2}		
10^{-3}		
10^{-4}		

(H p174, H+ p2)

See N5 for negative powers.

2 Write these numbers in standard form.

   a It takes 0.1 seconds to blink an eye.

   b A camera takes 0.008 seconds to take a picture.

   c It takes about 0.000 000 003 4 seconds for light to travel 1 metre.

   d One day is 86 400 seconds.

   e One year is 31 536 000 seconds.

   (H p176, 178, H+ p4, 6)

3 1.5×10^{-1} 1.05×10^{-1} 9.8×10^{-2} 8.9×10^{-2}

   a Rewrite the numbers as decimals.

   b Put the numbers in order of size, smallest first.

   (H p178, H+ p6)

4 The table shows the mass of three different subatomic particles.

Particle	Electron	Proton	Neutron
Mass (kg)	9.109×10^{-31}	1.673×10^{-27}	1.675×10^{-27}

Put the particles in order of mass, smallest first.

(H p178, H+ p6)

5 The weight of one paper clip is 4.3×10^{-1} g.

   Calculate the number of paper clips needed to weigh 1 kilogram.

   Give your answer to an appropriate degree of accuracy.

   (H p178, H+ p6)

6 The length of a paper clip is 33 mm.

   Paper clips are placed end to end in a line for 1×10^6 mm.

   Calculate the number of paper clips that are needed.

   Give your answer to an appropriate degree of accuracy.

   (H p176, H+ p4)

7 Using standard form, estimate the value of

    a $63\,000 \times 0.004$ **b** $\dfrac{21\,000}{0.049}$ **c** $\dfrac{300 \times 0.07}{500\,000}$

    (H+ p206)

8 The volume of a cube is $3.75 \times 10^6 \text{ cm}^3$.

    Calculate the length of one side, giving your answer in standard form.

See S4 for volume.

    (H p176, H+ p4)

9 The density of gold is $1.93 \times 10^4 \text{ kg/m}^3$.

    Calculate the mass of $1.5 \times 10^{-5} \text{ m}^3$ of gold.

    Give your answer in standard form.

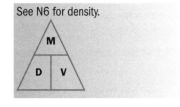

See N6 for density.

    (H p176, 178, H+ p4, 6)

10 It is estimated that the world population will be $8\,467\,000\,000$ by 2025.

    a Write this number in standard form.

    It is estimated that the population of China will be 1.488×10^9 by 2025.

    b Express the population of China as a percentage of the world population.

    (H p176, H+ p4)

11 The radius of the Moon is 1.7×10^3 km.

    a Use the formula $\text{Volume} = \frac{4}{3}\pi r^3$ where r is the radius, to calculate the volume of the Moon, giving your answer in standard form. State the units of your answer.

    The radius of Earth is 6.4×10^3 km.

    b Calculate the volume of Earth, giving your answer in standard form.

    c Calculate the ratio of the volumes, giving your answer in the form

    Volume of the Moon : Volume of Earth = 1 : ?

    Give your answer to a suitable degree of accuracy.

    (H p176, H+ p4)

12 $y^2 = \dfrac{ab}{a + b}$

    $a = 3 \times 10^8$

    $b = 2 \times 10^7$

    Find y.

    Give your answer in standard form correct to 2 significant figures.

    (Edexcel Ltd., 2003) 3 marks

Keywords
Expression
Factorisation
Indices
Simplify
Substitute
Term

- In algebra, you can use a letter to represent **any** number.
  If you have n full trays of 6 eggs each, you will have $6n$ eggs.

- An **expression** is a collection of **terms**.
  $4a + 3b + c^2$ is an expression.
  $4a$ is one term in the expression.

$4a$ means $4 \times a$

c^2 means $c \times c$

q means $1q$

- You **simplify** expressions involving addition and subtraction by collecting like terms.
  $4p + 3p + 2q - q = 7p + q$

- You can simplify expressions involving multiplication and division in algebra.
  $3 \times d = 3d \qquad 8 \times a \times b \times c = 8abc$
  $d \div 8 = \dfrac{d}{8} \qquad 6b \times 4b = 24b^2$

See N5 for squared and cubed.

- You use powers or **indices** to write repeated multiplication.
  $6^5 = 6 \times 6 \times 6 \times 6 \times 6$
  $b^5 = b \times b \times b \times b \times b$

The small raised number is the power or index.

- When you multiply terms of the same letter, you add the indices.
  $x^a \times x^b = x^{a+b}$
  $e^3 \times e^2 \times e^4 = e^{3+2+4} = e^9$

- When you divide terms of the same letter, you subtract the indices.
  $x^a \div x^b = x^{a-b}$
  $e^2 \div e^3 = e^{2-3} = e^{-1}$

$e^{-1} = \dfrac{1}{e}$

- You expand brackets by multiplying pairs of terms.
  $4\,(x + 2) = 4 \times x + 4 \times 2$

  $(p + 4)\,(p + 3) = p \times p + p \times 3 + 4 \times p + 4 \times 3$

  F   L
  O

F   Firsts
O   Outers
I   Inners
L   Lasts

- **Factorising** is the 'reverse' of expanding brackets.
  $8x - 14 = 2(4x - 7)$
  $x^2 - 9 = (x + 3)(x - 3)$
  $x^2 - 3x - 4 = (x + 1)(x - 4)$

Check the factorisation by expanding the brackets.

- You can **substitute** numbers into an expression and work out the value of the expression.
  If $a = 4$, $b = 3$, $c = 2$ then $6a^2b + c = 6 \times 4^2 \times 3 + 2 = 290$

See N3 for BIDMAS.

Example

The perimeter of a rectangle is $8a + 2b$.

The area of the rectangle is $4ab$.

Find the length and width of the rectangle.

width

length

See S4 for area and perimeter.

Half the perimeter $= \frac{1}{2}(8a + 2b)$

$$= \frac{1}{2} \times 8a + \frac{1}{2} \times 2b$$

$$= 4a + b$$

Length $\times$ width $= 4ab$
Length $+$ width $= 4a + b$ $\Bigg\}$ length $= 4a$, width $= b$

Example

Simplify

a $3x^2y \times 2xy^2$

b $(4y)^3$

c $30xy^2 \div 5x^2y$

a $3x^2y \times 2xy^2 = 3 \times 2 \times x^2 \times x \times y \times y^2 = 6x^3y^3$

b $(4y)^3 = 4y \times 4y \times 4y = 64y^3$

c $30xy^2 \div 5x^2y = \frac{30xy^2}{5x^2y} = \frac{\overset{6}{\cancel{30}} \times \cancel{x} \times \cancel{y} \times y}{\cancel{5} \times \cancel{x} \times x \times \cancel{y}} = \frac{6y}{x}$

Example

a Factorise fully **i** $x^2 + 6x - 16$

               ii $4x^2 - 16$

b Simplify $\dfrac{x^2 + 6x - 16}{4x^2 - 16}$

a **i** $x^2 + 6x - 16 = (x + 8)(x - 2)$

   ii $4x^2 - 16 = 4(x^2 - 4) = 4(x + 2)(x - 2)$

Factorise the numerator.
Factorise the denominator.
Cancel common terms.

b $\dfrac{x^2 + 6x - 16}{4x^2 - 16} = \dfrac{(x + 8)\cancel{(x - 2)}}{4(x + 2)\cancel{(x - 2)}} = \dfrac{(x + 8)}{4(x + 2)}$

Example

Simplify $\dfrac{1}{x - 2} - \dfrac{1}{x + 2}$

$\dfrac{1}{x - 2} - \dfrac{1}{x + 2} = \dfrac{(x + 2) - (x - 2)}{(x - 2)(x + 2)}$

$$= \dfrac{x + 2 - x + 2}{(x - 2)(x + 2)}$$

$$= \dfrac{4}{(x - 2)(x + 2)} = \dfrac{4}{(x^2 - 4)}$$

Exercise A1

1 A rectangle has a length of 2*a* and a width of *a*.

   a Calculate the perimeter of the rectangle.

   The rectangle is enlarged by scale factor 3.

   b Calculate the length and width of the enlarged rectangle.

   c Calculate the perimeter of the enlarged rectangle.

   (H p26, H+ p26)

See S6 for enlargements.

2 Claudia thinks of a number and then follows a set of instructions.

   a Copy and complete the table with *n* as Claudia's number.

Instructions	Algebra
Claudia's number	*n*
Add 4 to the number	
Double the answer	
Subtract the original number	
Subtract 10	
Subtract the original number	

   b What will Claudia get if she correctly follows the instructions?

   (H p26, 28, H+ p26)

3 Expand and simplify   **a** $3(x - 2) - 5(x - 3)$

                     b $4(2x - 3) - 2(3x + 5)$

                     c $(a + b)^2 + (a - b)^2$

   (H p28, H+ p28)

4 Factorise fully   **a** $x^2 - 8x$

                     b $x^2 - 16$

                     c $x^2 - 10x + 16$

   (H p32, 34, H+ p30, 32)

5 The length of a rectangle is $(2x + 1)$ cm.

   The width of the rectangle is $(x + 4)$ cm.

   The area of the rectangle is 72 cm².

   Show that $2x^2 + 9x - 68 = 0$.

   (H p30, H+ p28)

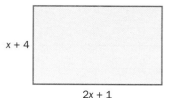

6 Simplify

  a $3h^4 \times 4h^3$

  b $8ab^4 \div 4a^2b$

  c $(5p^3)^2$

  (H+ p26)

7 Simplify

  a $\dfrac{x^2 - x}{x^2 - 1}$

  b $\dfrac{2x^2 + 3x + 1}{x^2 + 4x + 3}$

  c $\dfrac{x^2 - 9}{x^2 + 6x + 9}$

  (H+ p54)

8 Simplify

  a $\dfrac{1}{x} + \dfrac{1}{x + 1}$

  b $\dfrac{1}{x - 2} + \dfrac{1}{x - 3}$

  c $\dfrac{2}{x - 2} - \dfrac{3}{x + 3}$

  (H+ p56)

9 Simplify fully

  a $(3xy^2)^4$

  b $\dfrac{x^2 - 3x}{x^2 - 8x + 15}$

  (*Edexcel Ltd., 2005*) 5 marks

10 a Simplify $a^3 \times a^4$

  b Simplify $3x^2y \times 5xy^3$

  c Simplify $\dfrac{(x - 1)^2}{x - 1}$

  d Factorise $x^2 - 9$

  (*Edexcel Ltd., 2005*) 5 marks

Linear equations and inequalities

Keywords
Check
Equation
Inequality
Linear
Simultaneous
Solve
Substitute

● An **equation** links two or more expressions using an equals (=) sign.

$3x - 1 = 4x + 2$ is a **linear** equation.

● You can use a letter to represent particular values in an equation.

$8x - 4 = 36$ means x has to be 5.

Linear equations can be represented by straight lines.

● You can **solve** equations using the balance method.

An equation remains balanced if you do the same to both sides.

You can – add
– subtract
– multiply
– divide
– square root.

● If there are two unknowns in one equation, there is more than one solution.

● If there are two unknowns, you need two **simultaneous** equations to find the value of the unknowns.

The equations can be solved by algebra or by a graph.

The point of intersection of the straight lines gives the only solution.

$$\left. \begin{array}{l} x + y = 5 \\ -2x + y = 2 \end{array} \right\} x = 1, y = 4$$

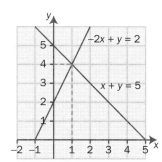

Simultaneous means at the same time.

See A6 for graphical solutions.

● When you solve an equation, you can **check** your answer by substitution.

You **substitute** the value back into the equation and the equation should balance.

Solving $3x + 4 = x - 6$ gives $x = -5$

Checking $3 \times -5 + 4 = -5 - 6$ ✓

● An **inequality** links two or more expressions using an inequality sign.

$2x + 7 \geqslant 9$ is an inequality.

Solving an inequality usually gives a range of values.

$<$ less than
$\leqslant$ less than or equal to
$>$ greater than
$\geqslant$ greater than or equal to

● When you multiply or divide by a negative number, you must reverse the inequality sign.

$-2x < 6$

Divide by -2 $x > -2$

● You can combine two inequalities.

$x > -1$ and $x \leqslant 2$
can be combined to give
$-1 < x \leqslant 2$

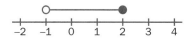

● can include 2
○ cannot include −1

Example

Solve $\dfrac{4 - x}{5} = x + 2$

$$\dfrac{4 - x}{5} = x + 2$$

Multiply both sides by 5. $(4 - x) = 5 \times (x + 2)$

Multiply out the brackets. $4 - x = 5x + 10$

Add x to both sides. $4 = 6x + 10$

Subtract 10 from both sides. $-6 = 6x$

Divide both sides by 6. $-1 = x$

Example

a   Find the range of values that satisfy both of these inequalities.

   $-2x \geqslant -2$ and $3x - 1 > -7$

b   Give the possible values of x if x is an integer.

a $-2x \geqslant -2$ $3x - 1 > -7$

   Divide both sides by -2. $x \leqslant 1$ Add 1 to both sides. $3x > -6$

                                         Divide both sides by 3. $x > -2$

● can include 1
○ cannot include -2

b $-1, 0, 1$

Example

Solve the simultaneous equations $4x - y = 3$ 1

                                                    $6x + 2y = 1$ 2

                                         $4x - y = 3$ 1

                                         $6x + 2y = 1$ 2

Equation 1 × 2 = Equation 3 $8x - 2y = 6$ 3

                                         $6x + 2y = 1$ 2

Equation 3 + equation 2 $14x = 7$

Divide both sides by 14. $14x = 7$

                                                $x = \dfrac{1}{2}$

Substitute in equation 1 $4 \times \dfrac{1}{2} - y = 3$

                                         $2 - y = 3$

                                                $y = -1$

Check in equation 2 $6 \times \dfrac{1}{2} + 2 \times -1 = 1$

                                         $3 + -2 = 1$ ✓

Exercise A2

1 Two of the angles in a parallelogram are $2a + 26°$ and $4a + 10°$.

   a Form and solve an equation in terms of a.

   b Calculate the four angles of the parallelogram.

   (H p54, H+ p50)

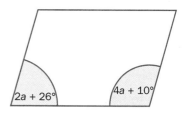
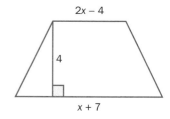

2 A trapezium is shown.

   All measurements are in centimetres.

   a Write the area of the trapezium in terms of x.

   The area of the trapezium is $36\,cm^2$.

   b Use this information to form an equation and to find the value of x.

   (H p54, H+ p50)

3 The areas of these two rectangles are the same.

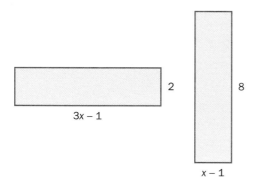

   a Derive and solve an equation that shows this information.

   b Calculate the area of each rectangle.

   (H p54, H+ p52)

4 a Solve the inequality $3x + 1 \geqslant 7$.

   b Write the equality shown by the diagram.

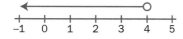

   c List the integers that satisfy both inequalities.

   (H p58, H+ p226)

5 a Solve the inequality $-5x < 15$.

   b Represent the solution on a number line.

   (H p58, H+ p226)

6 a Solve the inequalities $6x + 5 \leqslant 4x + 13$ and $2x > 1$.

b Give the possible values of x, if x is an integer, that satisfy both inequalities.

(H p58, H+ p226)

7 Solve

a $\dfrac{x + 3}{4} = \dfrac{x - 1}{3}$

b $\dfrac{5 - 2x}{3} = 3 - x$

c $\dfrac{(x + 2)}{4} + \dfrac{(4x - 3)}{10} = 8$

(H p56, H+ p58)

8 The equation of a straight line is $y = mx + c$.

A straight line goes through the points $(2, 11)$ and $(5, 23)$.

Form and solve two simultaneous equations to find the values of m and c.

(H p226, H+ p220)

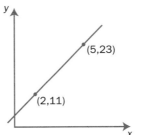

See A6 for $y = mx + c$.

9 Solve the simultaneous equations

$$3x + 5y = 7$$
$$5x - 3y = 6$$

(H p224, 226, H+ p220)

10 Solve the simultaneous equations

$$6x - 2y = 33$$
$$4x + 3y = 9$$

(*Edexcel Ltd., 2004*) 4 marks

11 a Solve $\dfrac{40 - x}{3} = 4 + x$

b Simplify $\dfrac{4x^2 - 6x}{4x^2 - 9}$

See A1 for factorising quadratics.

(*Edexcel Ltd., 2004*) 6 marks

Keywords
Factorisation
Quadratic
Simultaneous
Systematic
Trial and improvement

● A **quadratic** equation contains an x^2 term as the highest power.

$3x^2 - 2x + 5 = 0$ is a quadratic equation.

Quadratic equations have 0, 1 or 2 solutions.

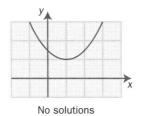 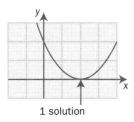 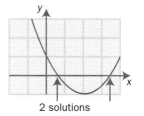

      No solutions   1 solution   2 solutions

You can determine a solution from a graph where it crosses an axis.

● Quadratic equations can be solved by

  ● **factorisation**

See A1 for factorisation.

  ● completing the square

  ● using the formula if $ax^2 + bx + c = 0$, $x = \dfrac{-b \pm \sqrt{b^2 - 4ac}}{2a}$

  ● graphically.

See A7 for graphical solutions.

● Some equations, for example, $x^3 + x = 10$ can be solved using **trial and improvement**.

In this method, you estimate a solution and substitute your estimate into the equation.

See N4 for trial and improvement.

If your estimate is not close enough, you improve it and try again.

You need to be **systematic** with your estimates.

● If there are two unknowns, you need two **simultaneous** equations to find the value of the unknowns.

The equations can be solved by algebra or by graph.

See A7 for graphical solutions.

The points of intersection of the straight line and the quadratic curve give the solutions.

$$\left.\begin{array}{l} y = x^2 - 4x - 5 \\ \\ y = x + 9 \end{array}\right\} \begin{array}{l} x = 7, y = 16 \\ \text{and} \\ x = -2, y = 7 \end{array}$$

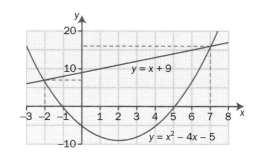

Example

Solve $x^2 - 3x - 28 = 0$

$x^2 - 3x - 28 = 0$

$(x - 7)(x + 4) = 0$

either $x - 7 = 0$ → $x = 7$

or $x + 4 = 0$ → $x = -4$

See A1 for factorising quadratics.

Example

a Express $x^2 + 6x + 5$ in the form $(x + a)^2 + b$, where a and b are integers.

b Hence or otherwise, solve $x^2 + 6x + 5 = 0$.

c What is the smallest value $x^2 + 6x + 5$ can have?

a $x^2 + 6x + 5 = (x + 3)^2 - 9 + 5$ as $(x + 3)^2 = x^2 + 6x + 9$

$\left. \begin{array}{l} = (x + 3)^2 - 4 \\ = (x + a)^2 + b \end{array} \right\}$ $a = 3, b = -4$

b $x^2 + 6x + 5 = 0$

$(x + 3)^2 - 4 = 0$

$(x + 3)^2 = 4$

$x + 3 = 2$ or $x + 3 = -2$

$x = -1$ or $x = -5$

c The smallest value of $(x + 3)^2 - 4$ is when $(x + 3) = 0$

The smallest value of $(x + 3)^2 - 4$ is $0^2 - 4 = -4$

Example

A rectangle has length $(x + 4)$ cm and width $(2x - 1)$ cm.

The area of the rectangle is 10 cm^2.

a Show that $2x^2 + 7x - 14 = 0$.

b Use the formula to find the values of x, correct to 2 decimal places.

c Calculate the length and width of the rectangle, correct to 2 decimal places.

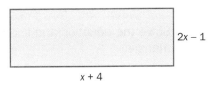

$2x - 1$

$x + 4$

a $(x + 4) \times (2x - 1) = 10$

$2x^2 - x + 8x - 4 = 10$

$2x^2 + 7x - 4 = 10$

$2x^2 + 7x - 14 = 0$

See A1 for FOIL.

b $ax^2 + bx + c = 0$ $a = 2, b = 7, c = -14$

$x = \dfrac{-b \pm \sqrt{b^2 - 4ac}}{2a}$ $= \dfrac{-7 \pm \sqrt{7^2 - 4 \times 2 \times -14}}{2 \times 2}$

$= \dfrac{-7 \pm \sqrt{49 + 112}}{4}$ $= \dfrac{-7 \pm \sqrt{161}}{4}$

$= \dfrac{-7 + 12.689}{4}$ or $\dfrac{-7 - 12.689}{4} = 1.422$ or -4.922

$= 1.42$ or -4.92 (to 2 dp)

c When $x = 1.422$, $2x - 1 = 2 \times 1.422 - 1 = 1.84$ cm (to 2 dp)

When $x = 1.422$, $x + 4 = 1.422 + 4 = 5.42$ cm (to 2 dp)

When $x = -4.922$, $2x - 1$ and $x + 4$ are negative and so are impossible for a rectangle.

43

Exercise A3

1 The square of a number is 1 more than the reciprocal of the number.

    a If the number is x, show that $x^3 - x = 1$.

    b One solution of $x^3 - x = 1$ is between 1 and 2.

       Find this value of x, correct to 1 decimal place.

(H p222)

> Reciprocal of x is $\frac{1}{x}$.

2 Factorise and solve

    a $x^2 + 11x + 24 = 0$

    b $2x^2 + 5x - 3 = 0$

    c $6x^2 - 5x + 1 = 0$

(H p220, H+ p104)

3 A right-angled triangle has lengths $x + 4$, $x + 2$ and x as shown.

    a Show that $x^2 - 4x - 12 = 0$.

> See S8 for Pythagoras.

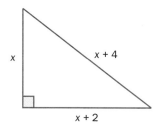

    b Solve the equation and find the lengths of the sides of the triangle.

(H p220, H+ p104, 218)

4 Use the formula to find the solutions to the quadratic equation

$$4x^2 - 3x - 2 = 0$$

Give your answers to 3 significant figures.

(H+ p106)

5 Solve the simultaneous equations

$$y = 4x^2$$
$$y = 8x - 3$$

(H+ p222)

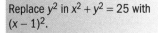

6 The equation of a circle is $x^2 + y^2 = 25$.

The equation of a straight line is $y = x - 1$.

The circle and the straight line intersect at two points.

a By eliminating y from the equations, show that $x^2 - x - 12 = 0$.

b Find the coordinates of the two points of intersection.

(H+ p224)

> Replace y^2 in $x^2 + y^2 = 25$ with $(x - 1)^2$.

7 The diagram below shows a 6-sided shape.

All the corners are right angles.

All measurements are given in centimetres.

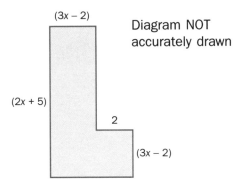

Diagram NOT accurately drawn

$(3x - 2)$

$(2x + 5)$

2

$(3x - 2)$

The area of the shape is $25\,cm^2$.

a Show that $6x^2 + 17x - 39 = 0$.

b i Solve the equation

$6x^2 + 17x - 39 = 0$

ii Hence work out the length of the longest side of the shape.

(*Edexcel Ltd., 2005*) 7 marks

8 A cuboid has a square base of side x cm.

The height of the cuboid is 1 cm more than the length x cm.

The volume of the cuboid is $230\,cm^3$.

a Show that $x^3 + x^2 = 230$.

The equation $x^3 + x^2 = 230$

has a solution between $x = 5$ and $x = 6$.

b Use a trial and improvement method to find this solution.

Give your answer correct to 1 decimal place.

You must show **all** your working.

(*Edexcel Ltd., 2003*) 6 marks

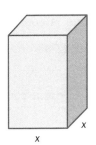

x

x

● A **formula** descibes the relationship between two or more **variables.**

It is usually an equation.

The formula for the volume of a cone is

$V = \frac{1}{3}\pi r^2 h$

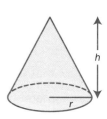

Keywords
Direct proportion
Formula
Inverse proportion
Subject
Substitute
Variable

V, r and h are the variables.
The values of r and h decide the value of V.

● You can **substitute** numbers into a formula.

If $A = \frac{1}{2}ab\sin C$ and $a = 6$, $b = 8$, $C = 30°$

then $A = \frac{1}{2} \times 6 \times 8 \times \sin 30°$

$\qquad = \frac{1}{2} \times 48 \times \frac{1}{2}$

$\qquad = 12$

$\sin 30° = \frac{1}{2}$

● The **subject** of the formula is the variable on its own on one side of the equals sign.

V is the subject of the formula
V = area of cross-section × length

● You can change the subject of a formula using the balance method for solving equations.

If $v = u + at$, then making t the subject, $t = \frac{v - u}{a}$

See A2 for the balance method.

● Two quantities are in **direct proportion** if the ratio between the quantities is constant.

In the graph, C is directly proportional to n.

$C \propto n$ or $C = kn$

where k is a constant.

$\propto$ means proportional.

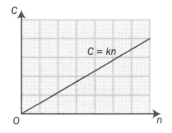

● When the product of two quantities is constant, the quantities are in **inverse proportion**.

In this graph, T is inversely proportional to n.

$T \propto \frac{1}{n}$ or $T = \frac{k}{n}$

where k is a constant.

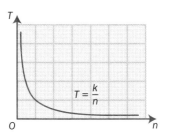

The formula linking f, u and v is

$$\frac{1}{f} = \frac{1}{u} + \frac{1}{v}$$

Calculate the value of f, if $u = 0.4$ and $v = 0.1$.

$\frac{1}{f} = \frac{1}{0.4} + \frac{1}{0.1}$	Substitute.
$\frac{1}{f} = 2.5 + 10$	Divide.
$\frac{1}{f} = 12.5$	Add.
$\frac{1}{12.5} = f$	Multiply 12.5 by f and then divide by 12.5.
$0.08 = f$	Divide.

The size of the exterior angle, e, of a regular polygon is inversely proportional to the number of exterior angles, n.

$e = 45°$ for a regular octagon.

a Express e in terms of n.

b Work out the value of e when $n = 6$.

c Calculate the number of sides of a polygon if $e = 20°$.

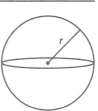

a $e \propto \dfrac{1}{n}$ $e = \dfrac{k}{n}$

           $45 = \dfrac{k}{8}$ Substitute.

           $k = 45 \times 8 = 360$ Find k.

           $e = \dfrac{360}{n}$ Substitute.

b When $n = 6$ $e = \dfrac{360}{6} = 60°$

c When $e = 20°$ $20 = \dfrac{360}{n}$

           $n = \dfrac{360}{20} = 18$ sides

The formula for the volume of a sphere is

$$V = \frac{4}{3}\pi r^3$$

where r is the radius of the sphere.

Rearrange the equation to make r the subject of the formula.

$V = \dfrac{4}{3}\pi r^3$

$3V = 4\pi r^3$

$\dfrac{3V}{4\pi} = r^3$

$\sqrt[3]{\dfrac{3V}{4\pi}} = r$ or $r = \sqrt[3]{\dfrac{3V}{4\pi}}$

Exercise A4

1 If the volume of a prism = area of cross-section × length,

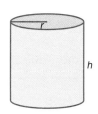

   a Derive the formula for the volume of a cylinder, where r is the radius of the circle and h is the length.

   b Calculate the volume of a cylinder, in terms of π, if $r = 5\,$cm and $h = 15\,$cm.

   State the units of your answer.

(H p184, H+ p182)

2 A circle of radius r fits inside a square as shown.

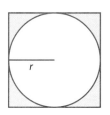

Show that the area of the shaded part, A, is given by the formula

$A = r^2(4 - \pi)$.

(H p184, H+ p182)

3 The area of a circle is given by

$$A = \pi r^2$$

Make r the subject of the formula.

(H p182, 186, H+ p184, 182)

4 The area of a trapezium is

$$A = \frac{1}{2}(a + b)h$$

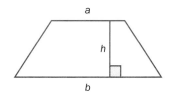

   a Make a the subject of the equation.

   b Calculate the value of a, if $A = 65\,$cm^2, $b = 18\,$cm and $h = 5\,$cm. State the units of your answer.

(H p182, 186, H+ p182, 184)

5 Rearrange each formula to make x the subject.

   a $8x + a = 5x + b$

   b $cx + a = dx + b$

   c $\dfrac{x + a}{x + b} = k$

(H p186, 188, H+ p184, 186)

6 The formula linking v, u, a and s, is $v^2 = u^2 + 2as$.

   a Make a the subject of the formula.

   b Calculate a, if $v = 5$, $u = 3$ and $s = 4$.

(H p182, 186, H+ p182, 184)

7 Three proportionality statements are shown.

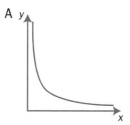 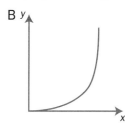 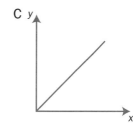

$y \propto x$ $y \propto \dfrac{1}{x}$ $y \propto x^2$

a Rewrite each statement as an equation.

b Match each graph with the proportionality statements.

A y

B y

C y

(H+ p302–308)

8 y is directly proportional to x^3.

The table shows some values of x and y.

x	2	4	8
y	12	90	768

One of the y values is wrong.

Showing all your working, decide which y value is wrong, and calculate the correct value.

(H+ p302, 304)

9 d is directly proportional to the square of t.

$d = 80$ when $t = 4$.

a Express d in terms of t.

b Work out the value of d when $t = 7$.

c Work out the positive value of t when $d = 45$.

(*Edexcel Ltd., 2005*) 6 marks

10 The shutter speed, S, of a camera varies inversely as the square of the aperture setting, f.

When $f = 8$, $S = 125$.

a Find a formula for S in terms of f.

b Hence, or otherwise, calculate the value of S when $f = 4$.

(*Edexcel Ltd., 2004*) 4 marks

Keywords
Counter-example
Difference
Linear
nth term
Position-to-term rule
Prove
Sequence
Term
Term-to-term rule

- The numbers in a **sequence** follow a pattern.

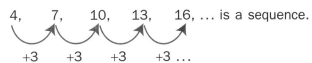

4,   7,   10,   13,   16, ... is a sequence.

+3   +3   +3   +3 ...

Each number in the sequence is called a **term**.

You find the **differences** by subtracting consecutive terms.

- A **term-to-term rule** links one term to the next term in the sequence.

> The differences are +3.

> The rule could use +,−, × and ÷

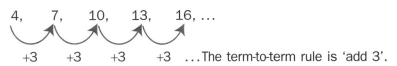

4,   7,   10,   13,   16, ...

+3   +3   +3   +3   ...The term-to-term rule is 'add 3'.

The term-to-term rule gives expressions for sequences of numbers.

$n, n + 1, n + 2, n + 3, n + 4, ...$ are consecutive numbers.

$2n, 2n + 2, 2n + 4, 2n + 6, ...$ are consecutive even numbers.

$2n − 1, 2n + 1, 2n + 3, 2n + 5, ...$ are consecutive odd numbers.

- A **position-to-term rule** links a term to its position in the sequence.

> The nth term is $3n + 1$.

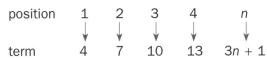

position	1	2	3	4	n
term	4	7	10	13	$3n + 1$

The position-to-term rule is often called the **nth term**.

The nth term allows you to calculate any term of the sequence.

The 100th term in this sequence is $3 \times 100 + 1 = 301$.

The differences for a **linear** sequence are equal.

The straight-line graph of $y = 3x + 1$ represents the sequence

4, 7, 10, 13, ...

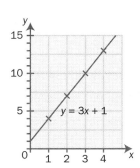

- Some sequences have special names.

2, 4, 6, 8, 10, ..., $2n$, ... are the even numbers.

1, 3, 5, 7, 9, ..., $2n − 1$, ... are the odd numbers.

1, 4, 9, 16, 25, ..., n^2, ... are the square numbers.

1, 3, 6, 10, 15, ..., $\frac{n(n+1)}{2}$, ... are the triangular numbers.

> Substitute $n = 1, 2, 3, 4, ...$ into the nth term to generate the sequence.

- You **prove** a statement by showing that it is true for **all** possible values of the variables.

You cannot choose particular values to test the statement.

You can show that a statement is false by finding a **counter-example**.

> You only need to find **one** counter-example.

The number of matchsticks in each pattern form a sequence.

Pattern 1   Pattern 2   Pattern 3   Pattern 4

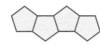

a State the number of matchsticks in each pattern.

b Write an expression, in terms of *n*, for the *n*th term.

c Use the pattern of the matchsticks to explain the *n*th term.

d Write a formula to link the number of matchsticks (*m*) and the pattern number (*n*).

e Calculate the number of matchsticks in Pattern 50.

a Pattern 1   Pattern 2   Pattern 3   Pattern 4

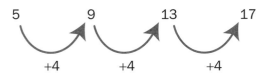

b *n*th term is $4n + 1$ as you add 4 matchsticks each time.

c   One matchstick is fixed and the pattern grows in steps of 4.

d $m = 4n + 1$

e When $n = 50$ $4n + 1 = 4 \times 50 + 1 = 201$

Any even number can be written as $2n$, where *n* is an integer.

a Write an expression for the first odd number after $2n$.

b Prove that the sum of three consecutive numbers, starting with an even number, is a multiple of 3.

a $2n + 1$

b Sum of three consecutive numbers, starting with an even number

$= 2n + (2n + 1) + (2n + 2)$

$= 6n + 3 = 3(2n + 1)$ which is divisible by 3.

Anne is asked to prove that the sum of **any** two numbers is always even. This is her proof. It is wrong.
Explain why and give the correct proof.

Even number = 2n
Another even number = 2n
Sum of even numbers = 2n + 2n = 4n
4n is divisible by 2 and so is even.

Anne assumes the even numbers are identical.

She should have used two different variables to represent the different even numbers.

Sum of even numbers $= 2r + 2s = 2(r + s)$

$2(r + s)$ is divisible by 2 and so is even.

Exercise A5

1 The nth term is given for these sequences.

Calculate the first five terms of each sequence.

a 10^n

b 2^n

c $(n + 1)(n + 3)$

(H p100, H+ p98)

See N5 for powers and indices.

2 Cubes are arranged as shown.

  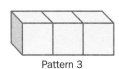

Pattern 1   Pattern 2   Pattern 3

a State the number of visible edges in each pattern.

b Write an expression, in terms of n, for the nth term.

c Use the visible edges of the cubes to explain the nth term.

d Write a formula to link the number of visible edges (v) and the pattern number (n).

e Calculate the pattern number if the number of visible edges is 104.

(H p102, 104, H+ p100)

3 **a** Maisie is investigating this pattern.

1st term	$1^2 + 1 \times 2 - 3$
2nd term	$2^2 + 2 \times 3 - 3$
3rd term	$3^2 + 3 \times 4 - 3$
4th term	$4^2 + 4 \times 5 - 3$
5th term	$5^2 + 5 \times 6 - 3$

Copy and complete the nth term for this pattern
nth term is $n^2 + \ldots$

b Jaya thinks the nth term is $(2n + 3)(n - 1)$.
Show that Jaya's expression for the nth term is the same as your answer for part **a**.

See A1 for FOIL.

(H p30, H+ p28)

4 The formula $p = 2n^2 + 1$ is suggested to generate prime numbers.

Find a counter-example to show this formula does not always give a prime number.

(H p190, H+ p188)

5 Alex claims that 'The square of a number is always larger than the number'. Find a counter-example to show this statement is false.

(H p190, H+ p188)

6 Find the nth term for this sequence.

1	2	3	4	5		n
↓	↓	↓	↓	↓		↓
$\frac{2}{5}$	$\frac{3}{7}$	$\frac{4}{9}$	$\frac{5}{11}$	$\frac{6}{13}$		?

Find the nth term for the numerator sequence and the nth term for the denominator sequence.

(H p102, H+ p100)

7 r is a positive integer.

a Write an expression, in terms of r, for the integer one more than r.

b Prove that the sum of the squares of any two consecutive integers is an odd number.

(H p190, H+ p188)

8 a $2r + 1$ is an odd number.

Write an expression, in terms of r, for the next odd number.

b Prove that the product of two consecutive odd numbers is three more than a multiple of 4.

(H p190, H+ p188)

9 Here are some patterns made from dots.

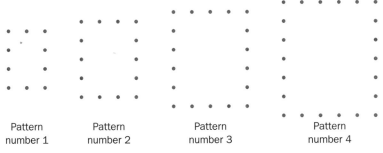

Pattern number 1   Pattern number 2   Pattern number 3   Pattern number 4

Write a formula for the number of dots, d, in terms of the pattern number, n.

(*Edexcel Ltd., 2004*) 2 marks

10 a Show that $(2a - 1)^2 - (2b - 1)^2 = 4(a - b)(a + b - 1)$

b Prove that the difference between the squares of **any** two odd numbers is a multiple of 8.

(You may assume that any odd number can be written in the form $2r - 1$, where r is an integer.)

(*Edexcel Ltd., 2003*) 6 marks

● A **linear** equation can be represented by a straight-line graph.

● The graph of the function $y = mx + c$ is a straight line, where m and c are numbers.

   m is the **gradient** or steepness.

   c is the **intercept** on the y-axis.

   $y = 2x - 1$ is a straight-line graph with a gradient of 2 and that passes through $(0, -1)$.

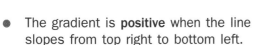

Keywords
Gradient
Intercept
Linear
Parallel
Perpendicular
Region

gradient $= \dfrac{\text{rise}}{\text{run}}$
$= \dfrac{2}{1}$
$= 2$

● The gradient is **positive** when the line slopes from top right to bottom left.

● The gradient is **negative** when the line slopes from top left to bottom right.

● **Parallel** lines have the same gradient.

   The straight lines $y = 3x$, $y = 3x - 1$, $y = 3x + 2$ all have a gradient of 3.

● The equation of a straight line can be rearranged.

   $5x + 3y = 15$ is the equation of a straight line.

See A4 for rearranging formulae.

● When two straight lines cross, the coordinates of the point of intersection represent the solution to both equations.

$\left.\begin{array}{l} y = 3x - 9 \\ y = x - 1 \end{array}\right\}x = 4, y = 3$

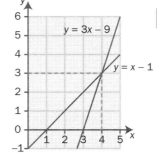

See A2 for simultaneous equations.

● The gradients of **perpendicular** lines are related.

$m \times \dfrac{-1}{m} = -1$

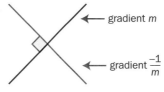

gradient m

gradient $\dfrac{-1}{m}$

● You can represent inequalities as **regions** on a graph.

   The unshaded region satisfies the three inequalities

   $x \geqslant -1$

   $y > x$

   $y < 2$

   Draw the lines $x = -1$

   $\qquad\qquad y = x$

   $\qquad\qquad y = 2$

   and then shade the regions.

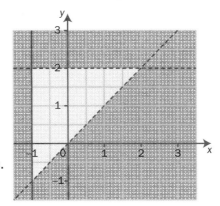

The solid line indicates x can be equal to -1.

A straight line **L** passes through the points $(-2, 3)$ and $(2, 1)$.

a Calculate the gradient of the line.

b Find the equation of the straight line.

c Find the equation of the straight line that is perpendicular to **L** and passes through the point $(2, 1)$.

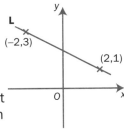

a gradient $= \dfrac{\text{rise}}{\text{run}} = \dfrac{-2}{4} = \dfrac{-1}{2}$

b
$$y = mx + c$$
$$y = -\tfrac{1}{2}x + c$$

When $x = 2$, $y = 1$ $1 = -\tfrac{1}{2} \times 2 + c$

$$1 = -1 + c$$
$$2 = c$$

Negative gradient

So the equation of **L** is $y = -\tfrac{1}{2}x + 2$.

c The gradient of a line perpendicular to **L** is 2.

The equation of a line perpendicular to **L** is $y = 2x + c$.

When $x = 2$, $y = 1$ $1 = 2 \times 2 + c$

$$1 = 4 + c$$
$$-3 = c$$

So the equation of the line perpendicular to **L** is $y = 2x - 3$.

$m \times \dfrac{-1}{m} = -1$

a On graph paper, shade the region of points whose coordinates satisfy the inequalities

$$x < 3, \quad y \leqslant 4, \quad x + y > 4$$

b Find the coordinates of the points that satisfy the inequalities, if x and y are integers.

a Draw the graphs $x = 3$

$$y = 4$$
$$x + y = 4$$

$x = 0$ $y = 4$

$x = 4$ $y = 0$

$x = 2$ $y = 2$

Shade the region. The solid line indicates y can be equal to 4.

b $(1, 4)$, $(2, 4)$ and $(2, 3)$

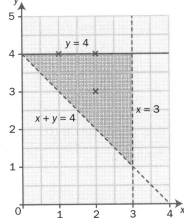

55

Exercise A6

1 **a** Draw the graphs of $3x + 2y = 16$

$$2x - y = 6$$

on the same grid for $0 \leqslant x \leqslant 6$.

b Give the coordinates of the point of intersection.

(H p122, H+ p122)

See A2 to solve simultaneous equations to check your answer.

2 The equation of a straight line is $3x + 5y = 15$.

a Rearrange the equation into the form $y = mx + c$.

b Hence or otherwise, state

    i the gradient of the line

    ii the coordinates of the point where the line crosses the y-axis.

(H p128, H+ p122)

3 Find the equation of a straight line that passes through $(2,0)$ and $(0,4)$.

Give your answer in the form $ax + by = c$, and state the values of a, b and c.

(H p130, H+ p124)

4 **a** On graph paper, shade the region of points whose coordinates satisfy the inequalities

$$x \geqslant 0, \quad y > 1, \quad x + y \leqslant 3$$

b Find the coordinates of the points that satisfy the inequalities, if x and y are integers.

(H+ p128, 130)

5

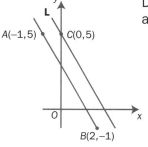

Diagram **NOT** accurately drawn

The diagram shows three points A $(-1,5)$, B $(2,-1)$ and C $(0,5)$.

The line **L** is parallel to AB and passes through C.

Find the equation of the line **L**.

(*Edexcel Ltd., 2005*) 4 marks

6   A line is drawn from $A(1,1)$ to $B(3,9)$.

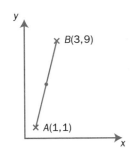

   a Calculate the gradient of the line *AB*.

   b Calculate the gradient of the line perpendicular to *AB*.

   c Calculate the coordinates of the midpoint of the line *AB*.

   d Find the equation of the line perpendicular to *AB* and passing through the midpoint of *AB*.

   (H+ p126, 194)

See S7 for the midpoint of a line segment.

7   **a** $-2 < x \leqslant 1$

   x is an integer.

   Write all the possible values of *x*.

   b $-2 < x \leqslant 1$ $y > -2$ $y < x + 1$

   x and *y* are integers.

   On a copy of the grid, mark with a cross (X), each of the six points which satisfy **all** three inequalities.

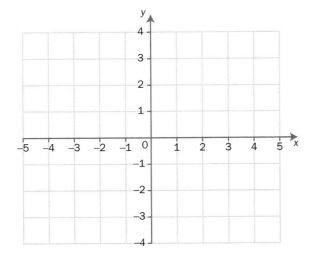

   (*Edexcel Ltd., 2003*) 5 marks

- The graph of a **quadratic** function is a parabola.

  It is a U-shaped curve.

  $y = x^2 + x$ is a quadratic equation.

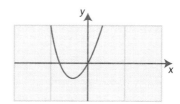

- The term with the highest power in a **cubic** equation is ax^3.

  $y = x^3 + 2x^2 - 11x - 12$ is a cubic equation.

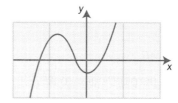

- A graph of a function involving a **reciprocal** is called a hyperbola.

  $y = \frac{1}{x}$ is a hyperbola.

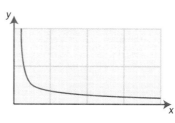

- An **exponential** graph has a term with one of its variables as an index.

  $y = 2^x$ is an exponential graph.

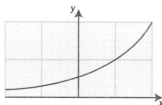

- The equation of a circle of radius r, with centre $(0, 0)$ is $x^2 + y^2 = r^2$.

  $x^2 + y^2 = 4$ is the equation of a circle.

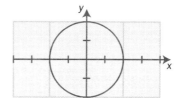

- The graphs of the sine and cosine ratios oscillate between two values.

  The maximum value is 1.

  The minimum value is -1.

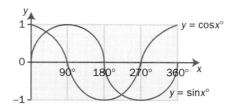

- When a curve and a straight line cross, the coordinates of the point(s) of intersection represent the solution(s) of the simultaneous equations.

  $y = x^2 - 4x - 5$
  $y = x + 9$
  $x = 7, y = 16$ and $x = -2, y = 7$

  See A3 for solving simultaneous equations by algebra.

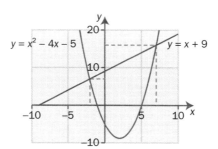

Example

a Draw the graph $y = x^2 - 2x - 3$ for $x = -3$ to $x = 5$.

b Use your graph to solve   **i** $x^2 - 2x - 3 = 0$

                                   ii $x^2 - 2x - 8 = 0$

c Draw the graph of $y = x - 3$ on the same grid.

The graphs of $y = x^2 - 2x - 3$ and $y = x - 3$ intersect at the points
A and B.

d Derive an equation, in the form $ax^2 + bx + c = 0$, that
represents the points A and B. State the values of a, b and c.

e State the values of x that satisfy this equation.

---

a

x	-3	-2	-1	0	1	2	3	4	5
y	12	5	0	-3	-4	-3	0	5	12

b i $0 = x^2 - 2x - 3$

      $y = x^2 - 2x - 3$

      Intersection of $y = 0$ and
      $y = x^2 - 2x - 3$

      $x = -1$ and $x = 3$

  ii $0 = x^2 - 2x - 8$

      $5 = x^2 - 2x - 3$

      $y = x^2 - 2x - 3$

      Intersection of $y = 5$
      and $y = x^2 - 2x - 3$

      $x = -2$ and $x = 4$

c $y = x - 3$

x	0	1	3
y	-3	0	-2

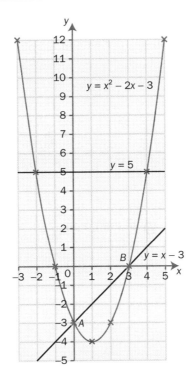

d $y = x^2 - 2x - 3$

    $y = x - 3$

    Substitute

    $x^2 - 2x - 3 = x - 3$

    $x^2 - 3x = 0$

    $x^2 + bx + c = 0$

    $a = 1$, $b = -3$, $c = 0$

e Using the graph, the intersection of $y = x^2 - 2x - 3$
    and $y = x - 3$ gives $x = 0$ and $x = 3$

Exercise A7

1 The graphs of $y = x^2 + x - 6$ and $y = 2x$ are shown in the diagram.

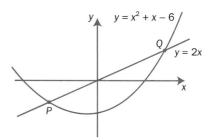

Find the coordinates of the points P and Q.

(H p298, H+ p222)

2 **a** Factorise $x^2 - x - 6$.

The graph $y = x^2 - x - 6$ crosses the x-axis at A and B.

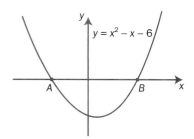

b Find the coordinates of A and B.

c Find the equation of the line of symmetry of the graph.

(H p296, H+ p294, 298)

3 **a** Copy and complete the table for $y = 2^x$.

x	−3	−2	−1	0	1	2	3
y			0.5				8

b On graph paper, draw the graph of $y = 2^x$.

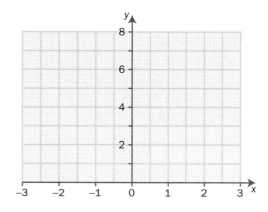

c Use your graph to solve the equations

   i $2^x = 3$ **ii** $2^x = 2x$

First draw the graph $y = 2x$.

(H+ p292, 296)

4 The straight line $y = x + 3$ intersects the circle $x^2 + y^2 = 9$ at the points A and B.

a Show that the x-coordinates of A and B can be found from the equation $x^2 + 3x = 0$.

b By solving $x^2 + 3x = 0$, find the coordinates of A and B.

(H+ p224, 298)

5 **a** Copy and complete this table of values for $y = x^3 + x - 2$.

x	−2	−1	0	1	2
y	−12			0	

b On a copy of the grid, draw the graph of $y = x^3 + x - 2$.

(*Edexcel Ltd., 2005*) 5 marks

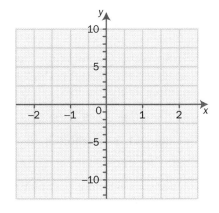

6 Diagram 1 is a sketch of part of the graph of $y = \sin x°$.

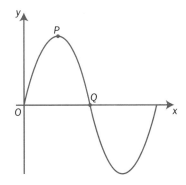

a Write the coordinates of

i P

ii Q

Diagram 2 is a sketch of part of the graph of $y = 3\cos 2x°$.

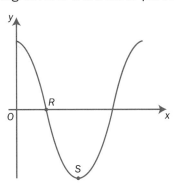

b Write the coordinates of

i R

ii S

(*Edexcel Ltd., 2005*) 4 marks

See A8 for transformations of functions.

- You can use mathematical functions to model real life.

- An exponential curve is often used to show population growth.

See A7 for the different functions.

Keywords
Acceleration
Distance—time graph
Gradient
Linear
Transform
Velocity—time graph

- Straight-line graphs can be used to represent quantities that have a linear relationship.

  $F = \frac{9}{5}C + 32$ is used to convert temperatures.

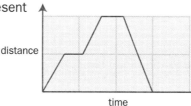

See A6 for $y = mx + c$

- A **distance**—**time** graph can represent a journey.

  The **gradient** of each line represents the speed.

  $\text{Speed} = \frac{\text{distance}}{\text{time}}$

  The steeper the gradient, the faster the speed.

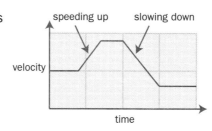

Time is always on the horizontal axis.

- A **velocity**—**time** graph describes a journey.

  The gradient of each line represents the **acceleration**.

  A horizontal line represents no acceleration, that is, a constant speed.

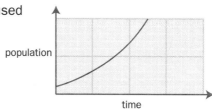

See N6 for speed.

See A6 for gradients.

- You can **transform** the graph of a function f(x).
  The equation of the function changes when the graph is transformed.

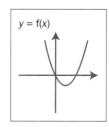

$y = f(x)$

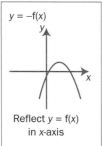

$y = f(x) + a$

Translate $y = f(x)$ by $\begin{pmatrix} 0 \\ a \end{pmatrix}$

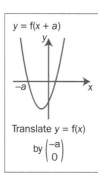

$y = f(x + a)$

Translate $y = f(x)$ by $\begin{pmatrix} -a \\ 0 \end{pmatrix}$

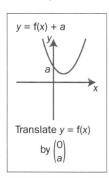

$y = af(x)$

Stretch $y = f(x)$ in y direction of a

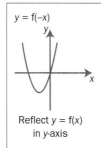

$y = -f(x)$

Reflect $y = f(x)$ in x-axis

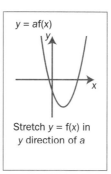

$y = f(-x)$

Reflect $y = f(x)$ in y-axis

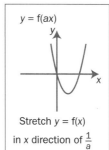

$y = f(ax)$

Stretch $y = f(x)$ in x direction of $\frac{1}{a}$

Example

The graph of $y = x^2$ is shown.

Sketch the graphs of

a $y = 2x^2$

b $y = (x + 2)^2$

c $y = -x^2$

d $y = x^2 + 5$

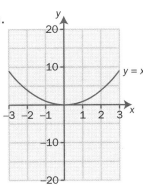

a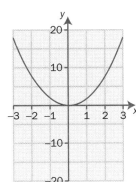

b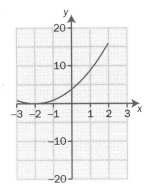

c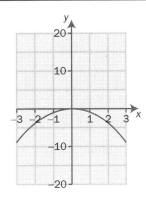

d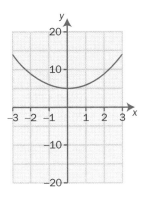

Example

The volume of a compost heap decreases each day.

The formula for the volume is

$V = kp^t$ where V is the volume in m^3

t is the time in days

k and p are positive constants.

After 2 days, the volume is $8\,m^3$.

After 3 days, the volume is $4\,m^3$.

Calculate

a the values of k and p

b the volume of the compost heap after 5 days.

a $V = kp^t$

1 $8 = kp^2$ When $t = 2$, $V = 8$

2 $4 = kp^3$ When $t = 3$, $V = 4$

   $0.5 = p$ Divide equation **2** by equation **1**

   $8 = k\,0.5^2 \rightarrow k = 32$ Substitute into equation **1**

So $V = 32 \times 0.5^t$

b When $t = 5$, $V = 32 \times 0.5^5$

                    $= 1\,m^3$

63

Exercise A8

1 Part of the shape of a tunnel entrance can be described by the quadratic equation

$$y = 5 - \frac{(x-6)^2}{8}$$

Calculate the coordinates of A, B and C.

(H p346)

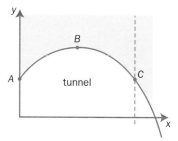

2 A parachutist jumps out of a moving plane.
The parachute jump has three stages.

0–5 seconds	Jumps out and free falls	Constant acceleration
5–10 seconds	The parachute opens	The speed slows down at a constant rate
10–20 seconds	The parachute resists the fall	The speed is constant

Choose the velocity—time graph that best describes the parachute jump.

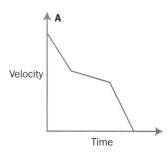 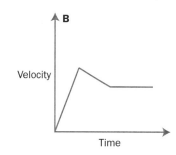 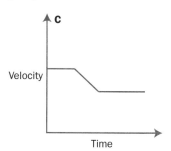

(H p338, 340)

3 The graph of $y = \cos x°$ is shown.

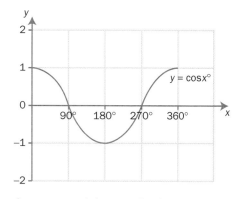

On a copy of the graph, draw the graphs of

a $y = 1 + \cos x°$

b $y = \cos 2x°$

c $y = -\cos x°$

(H+ p338–344)

4 The graph of $y = f(x)$ is shown on the grids.

  a On a copy of this grid, sketch the graph of $y = f(x - 1)$.

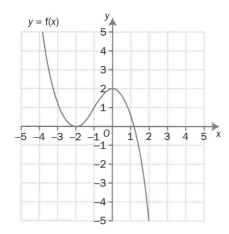

  b On a copy of this grid, sketch the graph of $y = 2f(x)$.

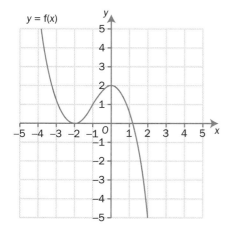

(*Edexcel Ltd., 2005*) 4 marks

5 Mr Patel has a car.
The value of the car on 1 January 2000 was £1600.
The value of the car on 1 January 2002 was £400.

The sketch graph shows how the value, £V, of the
car changes with time.

The equation of the sketch graph is $V = pq^t$
where t is the number of years after 1 January 2000, and
p and q are positive constants.

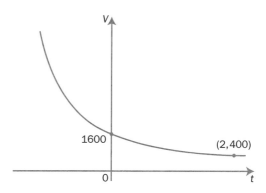

  a Use the information on the graph to
find the value of p and the value of q.

  b Using your values of p and q in the formula $V = pq^t$,
find the value of the car on 1 January 1998.

(*Edexcel Ltd., 2004*) 5 marks

• You need to know these angle properties.

Keywords
Alternate
Bearing
Corresponding
Parallel
Perpendicular
Polygon

360°

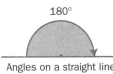

180°

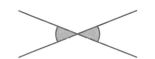

Angles at a point add to 360°.

Angles on a straight line add to 180°.

Vertically opposite angles are equal.

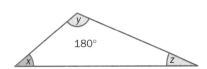

180°

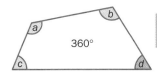

a b 360° c d

See S3 for types of triangles and quadrilaterals.

Angles in a triangle add to 180°.

Angles in a quadrilateral add to 360°.

• **Perpendicular** lines meet at right angles.

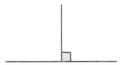

• **Parallel** lines are always the same distance apart.

You show parallel lines by sets of arrows.

Alternate angles are equal.

Corresponding angles are equal.

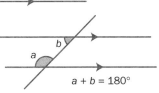

a b

a + b = 180°

Interior angles add to 180°.

• The exterior angles of any **polygon** add to 360°.

There are 5 equal angles of 72° for a regular pentagon.

For a regular polygon with n sides, each exterior angle = 360° ÷ n.

See S3 for regular polygons.

• The sum of the interior angles of any polygon depends on the number of sides.

A hexagon contains 4 triangles.

$4 \times 180° = 720°$

The sum of the interior angles of a hexagon is 720°.

For a regular polygon with n sides, the interior angle sum = $(n - 2) \times 180°$.

• You can give a direction using a **bearing**.

The bearing is measured clockwise from North and uses three figures.

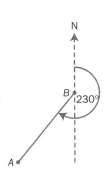

N

B 230°

A

The bearing of A from B is 230°.

The exterior angle of a regular polygon is half the interior angle.

Find the mathematical name of the polygon.

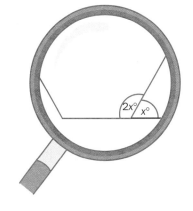

$2x + x = 180°$ Angles on a straight line add to 180°

$3x = 180°$

$x = 60°$ is the exterior angle.

The sum of the exterior angles = 360°

The number of exterior angles = 360° ÷ 60° = 6

The number of sides = 6, and so the polygon is a regular hexagon.

The diagram shows two towns A and B.

Calculate   **a**   the bearing of B from A

b   the bearing of A from B.

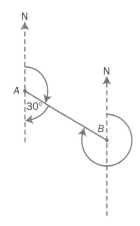

a   The bearing of B from A is
$180° − 30° = 150°$

b   The bearing of A from B is
$360° − 30° = 330°$

a   Calculate the sum of the interior angles of a heptagon.

b   Hence, or otherwise, calculate the value of $a°$.

a

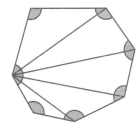

7 sides

5 triangles

$5 × 180° = 900°$

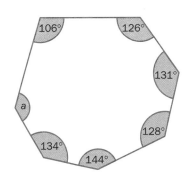

b $a° + 106° + 126° + 131° + 128° + 144° + 134° = 900°$

$a° + 769° = 900°$

$a° = 131°$

Exercise S1

1 The angles in a quadrilateral are
$7x + 6$, $7x + 1$, $4x - 4$ and $3x$.

Calculate the size of each angle in
the quadrilateral.

(H p88)

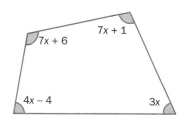

2 Calculate the size of the angles marked with letters.

a

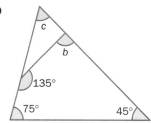

b

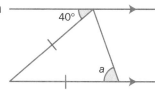

(H p86)

3 A regular polygon has 24 sides.

Calculate the size of   **a** an exterior angle

b an interior angle.

(H p88)

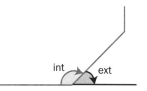

4 The exterior angles of a polygon are in the ratio $1:2:3:4:5$.

Calculate the size of each exterior angle.

(H p88)

See N6 for ratios.

5 Calculate the value of x in each diagram.

See A2 for solving equations.

a

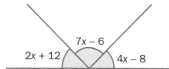

b

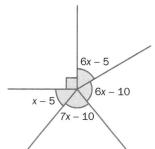

(H p85, H+ p85)

6 The diagram shows the positions of the
peaks of two mountains A and B.

Calculate **a** the bearing of B from A

b the bearing of A from B.

(H p242)

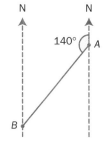

7 **a** Calculate the sum of the interior angles of a hexagon.

b Hence, or otherwise, calculate the size of each angle.

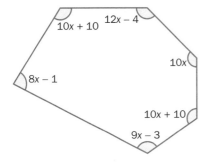

(H p88)

8 Work out the size of the angle marked *a*.

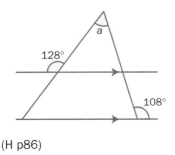

(H p86)

9 A lighthouse, *L*, is 3.2 km due West of a port, *P*.

A ship, *S*, is 1.9 km due North of the lighthouse, *L*.

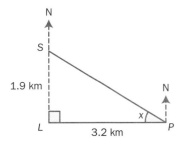

Diagram **NOT**
accurately drawn

a Calculate the size of the angle marked *x*.

Give your answer correct to 3 significant figures.

See S8 for trigonometry to calculate the angle *x*°.

b Find the bearing of the port, *P*, from the ship, *S*.

Give your answer correct to 3 significant figures.

(Edexcel Ltd., 2005) 4 marks

69

● You need to know these parts of a circle:
chord, **sector**, **segment**, arc and **tangent**.

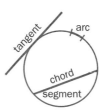

Keywords
Cyclic quadrilateral
Sector
Segment
Tangent

The longest chord is the diameter.

You need to know these angle properties
of a circle:

● The angle at the centre of a circle is double the
angle at the circumference from the same arc.

● Angles from the same arc in the same segment
are equal.

● The angle in a semicircle is a right angle.

● The opposite angles of a **cyclic quadrilateral**
add up 180°.

$x + y = 180°$

A cyclic quadrilateral has all four
vertices on the circumference of
a circle.

● The angle between the tangent and the radius
at a point is a right angle.

● Two tangents drawn from a point P to a circle
are equal in length.

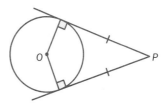

● The perpendicular line from the centre of
a circle to a chord bisects the chord.

● The angle between a tangent and a chord
through the point of contact is equal to the
angle in the alternate segment.

This is the alternate segment theorem.

Explain why *C* is not the centre of the circle.

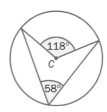

$\frac{1}{2}$ of 118° = 59° **not** 58°

Angle at *C* is **not** twice the angle at circumference.

Therefore *C* is not the centre of the circle.

A, *B* and *C* are points on the circumference of a circle.

AB is a diameter of the circle.

AC = 6 cm and *CB* = 8 cm.

a Find the value of angle *a*.

b Calculate the length of *AB*.

c Calculate the area of the circle, leaving your answer in terms of π.

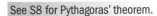

---

a *a* = 90°   Angle in a semicircle.

b $AB^2 = 6^2 + 8^2$ using Pythagoras' theorem as *ABC* is a
              right-angled triangle.

      = 36 + 64

      = 100

    AB = 10

c Area of the circle = $\pi \times 5^2 = 25\pi\,\text{cm}^2$ as the radius is 5 cm.

See S8 for Pythagoras' theorem.

See S4 for areas.

A, *B*, *C* and *D* are points on the circumference of the circle.

XAD and *XBC* are straight lines.

XA = *XB*

Show that *ABCD* is an isosceles trapezium.

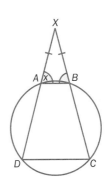

Let angle *XAB* = *x*

So angle *XBA* = *x*   Isosceles triangle

angle *BAD* = 180° − *x*   Angles on a straight line

angle *ABC* = 180° − *x*   Angles on a straight line

angle *ADC* = *x*   Opposite angles of a cyclic quadrilateral

angle *BCD* = *x*   Opposite angles of a cyclic quadrilateral

So *ABCD* is an isosceles trapezium, because of the angle properties.

71

Exercise S2

1 *A*, *B*, *C* and *D* are points on the circumference of the circle.

Angle *ABD* = 35°.

Angle *CBD* = 30°.

Angle *BCA* = 65°.

a Find the values of the angles *a*, *b*, *c* and *d*.

b Show that the opposite angles of a cyclic quadrilateral add to 180°.

(H p90, H+ p86)

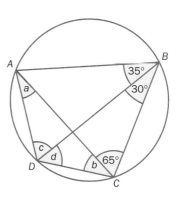

2 *A*, *B*, *C* and *D* are points on the circumference of the circle.

ABC is an isosceles triangle.

BA = *BC*.

Angle *ADE* = 108°.

Calculate the values of *x*, *y* and *z*.

(H p90, 94, H+ p86, 90)

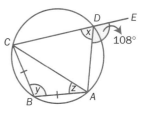

3 *A*, *B* and *C* are points on the circumference of the circle.

AC = 8 cm, *AB* = 15 cm and *CB* = 17 cm.

Show that *CB* passes through the centre of the circle.

(H p92, H+ p88)

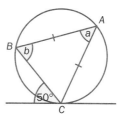

4 *A*, *B* and *C* are points on the circumference of the circle.

ABC is an isosceles triangle.

There is a tangent drawn through *C*.

The angle between the tangent and *BC* is 50°, as shown.

Calculate the values of *a* and *b*.

(H+ p92)

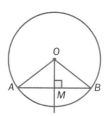

5 *O* is the centre of the circle.

AB is a chord.

A line is drawn from *O*, that is perpendicular to *AB* and crosses *AB* at *M*.

Prove that the triangles *OAM* and *OBM* are congruent.

(H+ p92, 158, 160)

See S7 for congruency.

6 Two circles intersect at *A* and *B*.

AC and AD are diameters.

Show that CBD is a straight line.

(H p92, H+ p88)

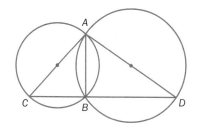

7

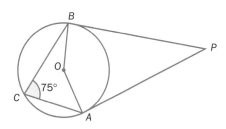

Diagram **NOT** accurately drawn

In the diagram, *A*, *B* and *C* are points on the circumference of a circle, centre *O*.

PA and PB are tangents to the circle.

Angle ACB = 75°.

a i Work out the size of angle AOB.

   ii Give a reason for your answer.

b Work out the size of angle APB.

(*Edexcel Ltd., 2005*) 5 marks

8

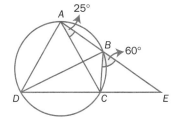

Diagram **NOT** accurately drawn

A, *B*, *C* and *D* are four points on the circumference of a circle.

ABE and DCE are straight lines.

Angle BAC = 25°.

Angle EBC = 60°.

a Find the size of angle ADC.

b Find the size of angle ADB.

Angle CAD = 65°.

Ben says that BD is a diameter of the circle.

c Is Ben correct? You must explain your answer.

(*Edexcel Ltd., 2004*) 4 marks

- A **polygon** is a 2-D shape with many sides and many angles.
  A **quadrilateral** is a polygon with 4 sides and 4 angles.

- You need to know the properties of these quadrilaterals.

Keywords
Plane of symmetry
Polygon
Prism
Quadrilateral
Regular
Symmetry

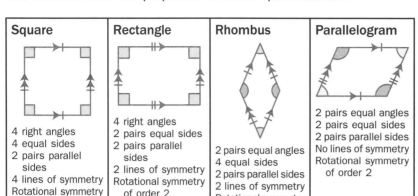

Square	Rectangle	Rhombus	Parallelogram
4 right angles 4 equal sides 2 pairs parallel sides 4 lines of symmetry Rotational symmetry of order 4	4 right angles 2 pairs equal sides 2 pairs parallel sides 2 lines of symmetry Rotational symmetry of order 2	2 pairs equal angles 4 equal sides 2 pairs parallel sides 2 lines of symmetry Rotational symmetry of order 2	2 pairs equal angles 2 pairs equal sides 2 pairs parallel sides No lines of symmetry Rotational symmetry of order 2

The angle sum of a quadrilateral is 360°.

The square is the only regular quadrilateral, as it has 4 equal sides and 4 equal angles.

Trapezium	Isosceles trapezium	Kite
1 pair parallel sides No lines of symmetry Rotational symmetry of order 1	2 pairs equal angles 1 pair equal sides 1 pair parallel sides 1 line of symmetry Rotational symmetry of order 1	1 pair equal angles 2 pairs equal sides No parallel sides 1 line of symmetry Rotational symmetry of order 1

Polygons

Sides	Name
3	triangle
4	quadrilateral
5	pentagon
6	hexagon
7	heptagon
8	octagon
9	nonagon
10	decagon

- A regular polygon with n sides has n lines of **symmetry**.

- A regular polygon with n sides has rotational symmetry of order n.

A **regular** shape has equal sides and equal angles.

The plural of vertex is vertices.

- A solid is a three-dimensional (3-D) shape.
  A **face** is a flat surface of a solid.
  An **edge** is the line where two faces meet.
  A **vertex** is a point at which three or more edges meet.

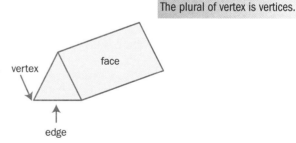

- A **prism** has a constant cross-section.
  This is a hexagonal prism.

You name a prism by the shape of its cross-section.

- A **plane of symmetry** divides a 3-D shape into two identical halves.
  A cuboid has 3 planes of symmetry.

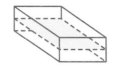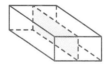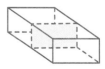

- You can draw a 3-D shape from different directions.
  The directions are plan, front elevation and side elevation.

See the example.

- A net is a 2-D shape that can be folded to form a 3-D shape.

This is the net of a solid.

a Draw a sketch of the solid.

b Give the mathematical name of the solid.

c State the number of edges of the solid.

d State the number of planes of symmetry of the solid.

A pyramid has faces that taper to a common point.
You name a pyramid by the shape of its base.

a

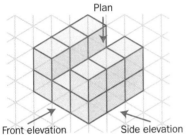

b tetrahedron or triangular-based pyramid

c 6 edges

d 6 planes of symmetry (one through each edge)

Draw the plan, front elevation and side elevation for the shape.

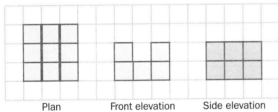

The 3-D shape is drawn on isometric paper.

Notice the bold line in the plan, when the level of the cubes alters.

Plan   Front elevation   Side elevation

A 3 by 4 by 5 cuboid is placed on 3-D coordinate axes as shown.

Write the coordinates of the points A, B and C in the form (x, y, z).

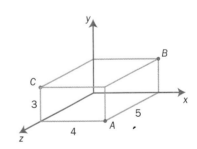

A is (4, 0, 5) because point A is 4 units along the x-axis
0 units along the y-axis
5 units along the z-axis.

B is (4, 3, 0) because point B is 4 units along the x-axis
3 units along the y-axis
0 units along the z-axis.

C is (0, 3, 5) because point C is 0 units along the x-axis
3 units along the y-axis
5 units along the z-axis.

Exercise S3

1 Draw a quadrilateral that has no lines of symmetry, but has rotational symmetry of order 2. Give its mathematical name.

(H p194)

2 A net consists of two equilateral triangles and three rectangles arranged as shown.

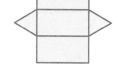

    a Draw a sketch of the solid and give its mathematical name.

    b Draw **one** plane of symmetry in your sketch and state the total number of planes of symmetry of the solid.

(H p194)

3 The points $A(-1, 2)$, $B(1, 1)$ and $C(0, -2)$ are marked on the grid.

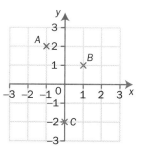

    a Give the two possible coordinates of another point D, such that $ABCD$ is a parallelogram.

    b Calculate the area of each parallelogram.

(H p16, 121)

4   **a** Name three quadrilaterals with perpendicular diagonals.

    b A square has diagonals of length $2x$.

    Calculate, in terms of x, the area of the square.

(H p196)

5 The plan, front elevation and side elevation are shown for a 3-D solid.

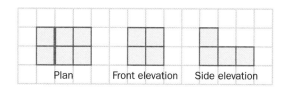

Draw the 3-D shape on isometric paper.

(H p278)

6 The plan, front elevation and side elevation are shown for a 3-D solid.

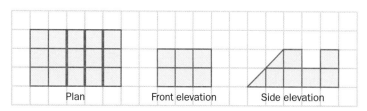

Draw a sketch of the 3-D solid.

(H p278)

7 The diagram shows a right-angled triangular prism.

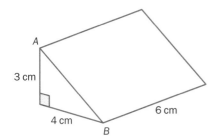

a Copy the diagram and draw the plane of symmetry for the prism.

b Use Pythagoras' theorem to calculate the distance *AB*.

See S8 for Pythagoras' theorem.

c On square grid paper draw an accurate net for the solid.

d Calculate the total surface area of the prism.

See S4 for surface area and volume.

e Calculate the volume of the prism.

(H p194)

8 A 2 by 2 by 4 cuboid is placed on 3-D coordinate axes as shown.

M is the midpoint of *AB*.

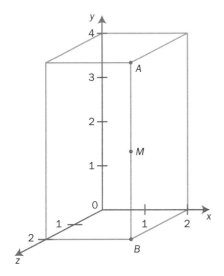

Write the coordinates of the points *A*, *B* and *M* in the form (x, y, z).

(H p202)

- The **perimeter** is the distance round a 2-D shape.
  It is a length (L).
  The perimeter of a circle is the circumference.
  $C = \pi d \quad C = 2\pi r$

  Arc length $= \dfrac{\theta}{360} \times$ circumference

- The **area** is the amount of surface a 2-D shape covers.
  It is length × length (L^2).
  You calculate area using these formulae.

You can measure perimeter in cm.

rectangle	parallelogram	triangle	triangle

Area $= l \times w$ Area $= b \times h$ Area $= \frac{1}{2}b \times h$ Area $= \frac{1}{2}ab\sin C$

You can measure area in cm².

trapezium	circle	sector

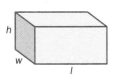 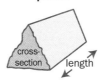

Area $= \frac{1}{2}(a + b)h$ Area $= \pi r^2$ Area $= \dfrac{\theta}{360} \times$ area of circle

- The **volume** is the amount of space inside a 3-D shape.
  It is length × length × length (L^3).
  You calculate volume using these formulae.

You can measure volume in cm³.

cuboid	prism	cylinder

Volume $= l \times w \times h$ Volume = area of cross-section × length   Volume $= \pi r^2 h$

pyramid	cone	sphere

Volume $= \frac{1}{3} \times$ area of base $\times h$ Volume $= \frac{1}{3}\pi r^2 h$ Volume $= \frac{4}{3}\pi r^3$

- You can use dimensions to analyse expressions and formulae.
  $4\pi r^2$ has dimensions (length)^2 = L^2 = area

4 and π are numbers and have no dimensions.

- The **surface area** of a 3-D shape is the total area of the outside face(s), or the area of the net of the shape.

cylinder	cone	sphere

Curved surface area $= 2\pi rh$ Curved surface area $= \pi rl$ Surface area $= 4\pi r^2$

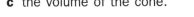

Example

Calculate **a** the vertical height h

       b the total surface area

       c the volume of the cone.

Leave your answers in terms of π if necessary.

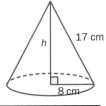

a Use Pythagoras' theorem $h^2 = 17^2 - 8^2 = 289 - 64$

                $h^2 = 225$ $h = 15\,\text{cm}$

See S8 for Pythagoras.

b Surface area $= \pi r l + \pi r^2 = \pi \times 8 \times 17 + \pi \times 8^2$

             $= \pi \times 136 + \pi \times 64$

             $= 200\pi\,\text{cm}^2$

c Volume $= \frac{1}{3}\pi r^2 h$

             $= \frac{1}{3}\pi \times 8^2 \times 15$

             $= 320\pi\,\text{cm}^3$

Example

Change $8\,\text{cm}^3$ to mm^3.

    $1\,\text{cm}^3$ $1000\,\text{mm}^3$

$1\,\text{cm}^3 = 1000\,\text{mm}^3$

$8\,\text{cm}^3 = 8 \times 1000\,\text{mm}^3 = 8000\,\text{mm}^3$

Example

A large cone has a height of h and a radius of r.

A smaller cone of height $\frac{h}{2}$ is cut off the top of the large cone, to leave a frustum.

Show that the volume of the frustum is $\frac{7\pi r^2 h}{24}$.

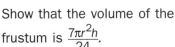

Volume of the large cone $= \frac{1}{3}\pi r^2 h$

Using similar triangles, the radius of the smaller cone is $\frac{r}{2}$

See S6 for similar triangles.

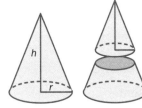

Volume of the small cone $= \frac{1}{3}\pi\left(\frac{r}{2}\right)^2 \frac{h}{2}$ $= \frac{\pi r^2 h}{24}$

Volume of the frustum $= \frac{1}{3}\pi r^2 h - \frac{\pi r^2 h}{24} = \frac{7\pi r^2 h}{24}$

Exercise S4

1 *A*, *B* and *C* are points on the circumference of a circle, centre *O*.

$AC = 12$ cm and $BC = 16$ cm.

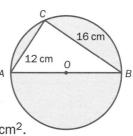

a State the value of angle *ACB*.

b Calculate the length of the diameter *AB*.

c Show that the shaded area is $4(25\pi - 24)$ cm^2.

(H p14, 18)

See S2 for circle properties.

See S8 for Pythagoras' theorem.

2 A prism has a cross-section of a trapezium.

The volume of the prism is 500 cm^3.

Calculate the value of *h*.

State the units of your answer.

(H p16, 280)

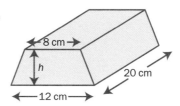

3 Change 5000 cm^2 to m^2.

(H p284, H+ p278)

4 Three cylinders are joined together as shown. *r* and *h* are lengths.

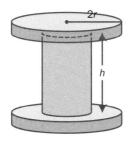

πr^2h^2 $3\pi r^2h$ $14\pi r^2 + 4\pi rh$ $8\pi r + 3\pi h$

Which of these expressions could represent

a the surface area

b the volume?

(H p284, H+ p278)

5 The diagram consists of a sector *OAB* of a circle and an isosceles triangle.

The centre of the circle is *O*.

The radius of the circle is 6 cm and the angle of the sector is 30°.

Show that the shaded area is $3(\pi - 3)$ cm^2.

(H+ p280)

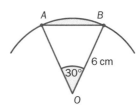

6

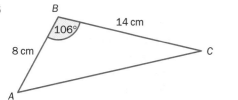

Diagram **NOT**
accurately drawn

ABC is a triangle.
AB = 8 cm.
BC = 14 cm.
Angle *ABC* = 106°.

Calculate the area of the triangle.
Give your answer correct to 3 significant figures.

(*Edexcel Ltd., 2005*) 3 marks

7

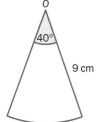

Diagram **NOT**
accurately drawn

The diagram shows a sector of a circle, centre *O*.
The radius of the circle is 9 cm.
The angle at the centre of the circle is 40°.

Find the perimeter of the sector.
Leave your answer in terms of π.

(*Edexcel Ltd., 2003*) 4 marks

8

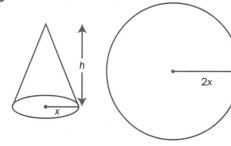

Diagram **NOT**
accurately drawn

The radius of the base of a cone is *x* cm and its height is *h* cm.
The radius of a sphere is 2*x* cm.
The volume of the cone and the volume of the sphere are equal.

Express *h* in terms of *x*.
Give your answer in its simplest form.

(*Edexcel Ltd., 2005*) 3 marks

- A **transformation** can change the size and position of a shape.

  You transform the object to the image.

- A **reflection** flips the shape over a **mirror line**.

  Each point in the image is the same distance from the mirror line as the corresponding point in the object.

  You describe a reflection by specifying the mirror line.

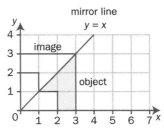

- A **rotation** turns a shape.

  You describe a rotation by giving

  — the **centre of rotation**

  — the angle of rotation

  — the direction of turn, either clockwise or anticlockwise.

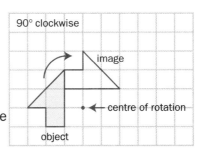

- A **translation** is a sliding movement.

  You describe a translation using a column vector $\begin{pmatrix} a \\ b \end{pmatrix}$

  which means a units in the x direction

  and $\qquad$ b units in the y direction.

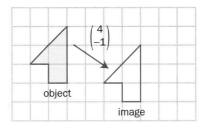

- In these transformations, the object and the image are **congruent**.

  In congruent shapes

  — corresponding angles are equal

  — corresponding sides are equal.

  See S7 for congruent shapes.

- A **vector** has both size and direction.

  A displacement, which has a fixed length in a fixed direction, can be represented by a vector.

  Vectors represented by parallel lines are multiples of each other.

  $\begin{pmatrix} 3 \\ 2 \end{pmatrix} =$

  a and **2a** are parallel vectors.

  The resultant vector completes the triangle of vectors.

  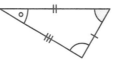

  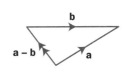
  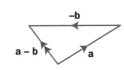

  Vector addition is commutative.

  a + **b** = **b** + **a**

  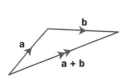
  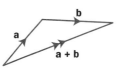
  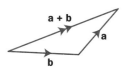

Example

Describe fully the transformation that maps

a triangle **A** to triangle **B**

b triangle **C** to triangle **B**

c triangle **C** to triangle **D**

d triangle **A** to triangle **D**.

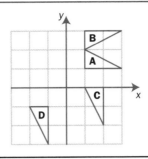

---

a Reflection in the line $y = 2$

b Rotation of 90° anticlockwise about $(0,1)$

c Translation of $\begin{pmatrix} -3 \\ -1 \end{pmatrix}$

d Reflection in the line $y = -x$

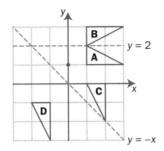

Example

ABC is a triangle.

X and Y are the midpoints of BA and BC respectively.

$\overrightarrow{BA} = 2\mathbf{a}$ and $\overrightarrow{BC} = 2\mathbf{b}$.

a Find, in terms of $\mathbf{a}$ and $\mathbf{b}$, vector representations for $\overrightarrow{AC}$ and $\overrightarrow{XY}$.

b Show that AC is parallel to XY.

c Find the value of k, such that $\overrightarrow{AC} = k\overrightarrow{XY}$.

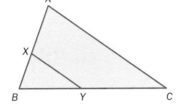

---

a $\overrightarrow{AC} = \overrightarrow{AB} + \overrightarrow{BC}$ using triangle ABC

   $= -2\mathbf{a} + 2\mathbf{b}$

   $= 2(-\mathbf{a} + \mathbf{b})$

   $\overrightarrow{BX} = \mathbf{a}$ and $\overrightarrow{BY} = \mathbf{b}$ as X and Y are midpoints

   $\overrightarrow{XY} = \overrightarrow{XB} + \overrightarrow{BY}$ using triangle XBY

   $= -\mathbf{a} + \mathbf{b}$

b AC is parallel to XY as $\overrightarrow{AC}$ is a multiple of $\overrightarrow{XY}$

c So $\overrightarrow{AC} = 2(-\mathbf{a} + \mathbf{b}) = 2\overrightarrow{XY}$ as $\overrightarrow{XY} = -\mathbf{a} + \mathbf{b}$

   but $\overrightarrow{AC} = k\overrightarrow{XY}$ and so $k = 2$

Exercise S5

1   Describe fully the transformation that maps

   a trapezium **A** to trapezium **B**

   b trapezium **B** to trapezium **C**

   c trapezium **A** to trapezium **D**

   d trapezium **C** to trapezium **E**

   e trapezium **E** to trapezium **B**.

   (H p164)

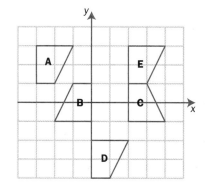

2   Copy the diagram.

   a Draw the reflection of triangle **A** in the line $y = x$.

   Label the new triangle as **B**.

   b Draw triangle **B** after a rotation of 90° clockwise about (0,0).

   Label the new triangle as **C**.

   c Describe fully the transformation that maps triangle **A** to triangle **C**.

   (H p166)

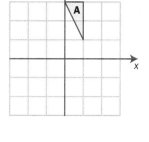

3   $ABCD$ is a trapezium, such that $AD = 2BC$.

   $\overrightarrow{BC} = \mathbf{a}$ and $\overrightarrow{AB} = \mathbf{b}$.

   Find, in terms of **a** and **b**, vector representations

   for **a** $\overrightarrow{AD}$

         b $\overrightarrow{AC}$

         c $\overrightarrow{CD}$

         d $\overrightarrow{BD}$

   (H+ p322)

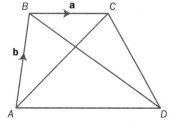

4   $ABCD$ is a parallelogram.

   The diagonals AC and BD meet at X.

   M and N are the midpoints of AB and DC respectively.

   $\overrightarrow{AB} = \mathbf{a}$ and $\overrightarrow{AD} = \mathbf{b}$.

   Find, in terms of **a** and **b**, vector representations

   for **a** $\overrightarrow{BD}$

         b $\overrightarrow{DC}$

         c $\overrightarrow{AC}$

         d $\overrightarrow{MX}$

         e $\overrightarrow{XN}$

   Hence or otherwise, show that M, X and N are on the same straight line.

   (H+ p322)

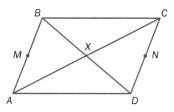

5 Triangle **A** and triangle **B** have been drawn on the grid.

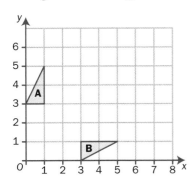

Describe fully the single transformation which will map triangle **A** onto triangle **B**.

(Edexcel Ltd., 2005) 2 marks

6

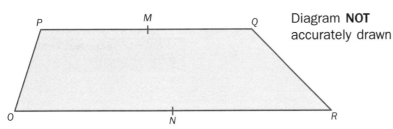

Diagram **NOT** accurately drawn

OPQR is a trapezium with *PQ* parallel to *OR*.

$\overrightarrow{OP} = 2\mathbf{b}$

$\overrightarrow{PQ} = 2\mathbf{a}$

$\overrightarrow{OR} = 6\mathbf{a}$

M is the midpoint of *PQ* and N is the midpoint of *OR*.

a Find the vector $\overrightarrow{MN}$ in terms of **a** and **b**.

X is the midpoint of *MN* and Y is the midpoint of *QR*.

b Prove that *XY* is parallel to *OR*.

(Edexcel Ltd., 2005) 4 marks

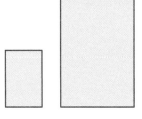

- **Similar** shapes have the same appearance but are different in size.

  One of the shapes is an enlargement of another shape.

Keywords
Centre of enlargement
Enlargement
Multiplier
Scale factor
Similar

Any two circles are similar.
Any two squares are similar.

- In an **enlargement,** the lengths change by the same **scale factor.**

  The scale factor is the **multiplier** of the lengths.

- In an enlargement,

  — corresponding angles are equal

  — corresponding lengths increase in the same ratio.

- You describe an enlargement by giving

  — the scale factor

  — the **centre of enlargement.**

The centre of enlargement fixes the position of the image.

- You use corresponding lengths to calculate the scale factor.

$$\text{Scale factor} = \frac{\text{length of the image}}{\text{length of the object}}$$

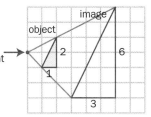

Scale factor 3
$$\frac{6}{2} = 3$$
$$\frac{3}{1} = 3$$

- An enlargement by a scale factor of less than 1 produces a smaller image.

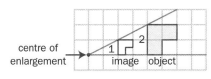

Scale factor $\frac{1}{2}$
$$\frac{1}{2} = \frac{1}{2}$$

- An enlargement by a negative scale factor produces an inverted image.

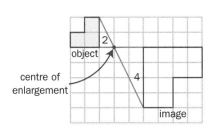

Scale factor −2
$$\frac{4}{2} = 2, \text{ but inverted image.}$$

- In maps and scale drawings, the lengths of lines and shapes are reduced or enlarged in proportion.

  The scale factor can be written as a ratio for maps and scale drawings.

  1 : 50 000 is an enlargement of scale factor 50 000.

See N6 for map scales.

- If the scale factor for length is L, then

  — the multiplier for length is L

  — the multiplier for area is $L \times L = L^2$

  — the multiplier for volume is $L \times L \times L = L^3$.

Example

ABCD is a trapezium.

D has coordinates (3,1).

The enlargement of *ABCD* is shown.

a Find the scale factor and the centre of enlargement.

b Give the coordinates of the point *D* after the enlargement.

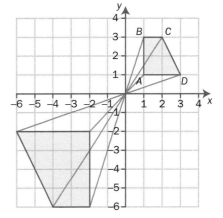

a Scale factor is −2

Centre of enlargement is (0,0)

b (−6,−2)

BE is parallel to *CD*.
AB = 8 cm
BC = 4 cm
CD = 9 cm
AE = 10 cm

Calculate *ED* and *BE*.

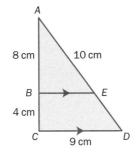

There are two similar triangles *ABE* and *ACD*.

$\frac{8}{12}=\frac{BE}{9}$ and so $BE = \frac{72}{12} = 6$ cm

$\frac{12}{8}=\frac{AD}{10}$ and so $AD = \frac{120}{8} = 15$ cm

So *ED* = 15 − 10 = 5 cm

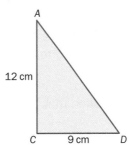

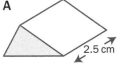

Prism **A** and prism **B** are mathematically similar.

A 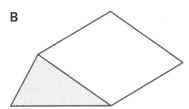 **B**

In prism **A**, the shaded area is 4.5 cm².

In prism **B**, the shaded area is 40.5 cm².

The length of prism **A** is 2.5 cm.

Calculate **a** the length of prism **B**

b the ratio of the volume of prism **A** to the volume of prism **B**, in the form 1:*n*, where *n* is an integer.

a Ratio of the areas is 4.5 : 40.5 or 1 : 9

Ratio of the lengths is 1 : 3

So length of prism **B** is 2.5 × 3 = 7.5 cm.

b Ratio of the volumes is 1 : 3³ = 1 : 27

$\sqrt{9}=3$

Exercise S6

1 A3-sized paper measures 297 mm by 420 mm.

A4-sized paper measures 210 mm by 297 mm.

Show that the two sizes of paper are mathematically similar to a specified degree of accuracy.

State this degree of accuracy.

(H p320, H+ p164)

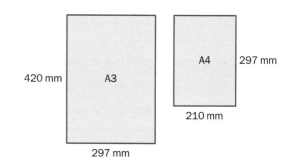

2 *AB* is parallel to *ED*.
$AB = 9.6$ cm
$ED = 12$ cm
$AC = 8$ cm
$CE = 11$ cm

a Show that the triangle *ABC* is similar to triangle *DEC*.

b Calculate the lengths *CB* and *CD*.

(H p322, H+ p164)

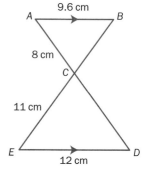

3 a On square grid paper, draw the enlargement of shape **A**, scale factor $-\frac{1}{2}$, using **O** as the centre of enlargement.

Label the enlarged shape as **B**.

b Describe fully the transformation that maps shape **B** to shape **A**.

(H + p162)

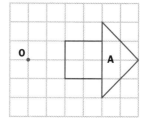

4 Cone **A** and cone **B** are mathematically similar.

The volume of cone **A** is 4687.5 cm^3.

The volume of cone **B** is 300 cm^3.

a Calculate the ratio of the volume of cone **A** to the volume of cone **B**, in the form $n:1$.

b Calculate the ratio of the area of the base of cone **A** to the area of the base of cone **B**, in the form $n:1$.

The vertical height of cone **A** is 15 cm.

c Calculate the vertical height of cone **B**.

(H + p166)

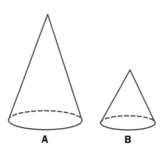

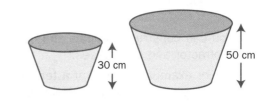

5 Two similar buckets are shown.

The height of the smaller bucket is 30 cm.

The height of the larger bucket is 50 cm.

The smaller bucket holds 13.5 litres of water.

Calculate the capacity of the larger bucket.

(H+ p166)

6

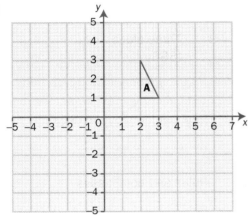

Copy the diagram.

Enlarge triangle **A** by scale factor $-1\frac{1}{2}$, centre O.

(*Edexcel Ltd., 2003*) 3 marks

7

4 cm

Diagram **NOT**
accurately drawn

Two cylinders, **P** and **Q**, are mathematically similar.

The total surface area of cylinder **P** is 90π cm².

The total surface area of cylinder **Q** is 810π cm².

The length of cylinder **P** is 4 cm.

a Work out the length of cylinder **Q**.

The volume of cylinder **P** is 100π cm³.

b Work out the volume of cylinder **Q**.

Give your answer as a multiple of π.

(*Edexcel Ltd., 2005*) 5 marks

- You can construct triangles and other 2-D shapes using a ruler, protractor and compasses.

    For example, the net of a tetrahedron can be constructed using just a ruler and compasses.

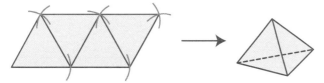

Keywords
Bisect
Congruent
Hypotenuse
Locus
Perpendicular

See S3 for nets.

You can construct a 60° angle using this method.

- You will always construct a unique triangle if you are given

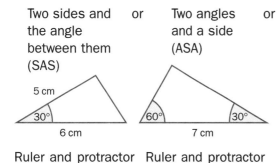

| Two sides and the angle between them (SAS) | or | Two angles and a side (ASA) | or | Right angle, the hypotenuse and a side (RHS) | or | Three sides (SSS) |

5 cm   30°   6 cm

60°   7 cm   30°

5 cm   3 cm

6 cm   8 cm   10 cm

Ruler and protractor   Ruler and protractor   Ruler, protractor and compasses   Ruler and compasses

- You can prove two triangles are **congruent** if one of these conditions is satisfied: SAS, ASA, RHS, SSS.

The longest side of a right-angled triangle is called the **hypotenuse**.

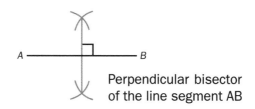

3 cm   8 cm   8 cm   3 cm

These two triangles are congruent (RHS).

- You can **bisect** angles and lines using a ruler and compasses.

Bisect means cut into two equal parts.

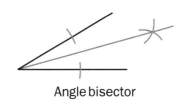

Angle bisector

A ———————— B

Perpendicular bisector of the line segment AB

- You can construct the **perpendicular** from a point *P* on the line.

See the example.

P

- You can construct the perpendicular from a point *P* to the line.

See the example.

P

- A **locus** is the path traced out by a moving point, which moves according to a rule.

    The locus of the points that are equidistant from *A* and *B* is the perpendicular bisector of the line *AB*.

A ———————— B

Construct the perpendicular from the point P on the line segment AB.

Construct the perpendicular from the point P to the line segment AB.

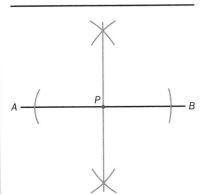

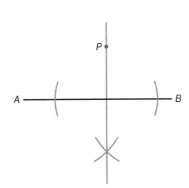

The points A and B are such that AB = 4 cm.

a   Construct the perpendicular bisector of AB.

b   Shade the region that satisfies the two conditions

  - further from A than B

  - less than 3 cm from A.

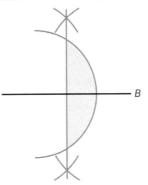

a   This is the red vertical line.

b   Set your compasses at 3 cm. Place at point A and draw an arc. The shaded region is on the opposite side of the perpendicular and within the arc.

AB is a chord of the circle with centre O.

The line OP is perpendicular to AB.

a   Prove that the triangles OAP and OBP are congruent.

b   Hence, show that P is the midpoint of AB.

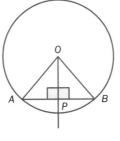

a   AO = OB (radii of the circle)

Angle OPA = angle OPB (OP is perp to AB)

OP is the same line in triangles OAP and OBP.

So triangles OAP and OBP are congruent (SAS).

b   Since triangles OAP and OBP are congruent, AP = PB.

So P is the midpoint of AB.

Exercise S7

1 The net of a 3-D shape consists of four equilateral triangles.

The length of one side of each triangle is 5 cm.

a State the mathematical name of the 3-D shape.

b Use a ruler and compasses only to accurately construct the net.

(H p244)

See S3 for nets of 3-D shapes.

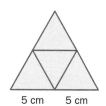

5 cm   5 cm

2 Two rotating water sprinklers are 25 metres apart.

One sprinkler sprays water within a 15 metre radius.

The other sprinkler sprays water within a 17.5 metre radius.

Draw a diagram to show the region that is sprayed by both sprinklers.

Use a scale of 1 cm to represent 5 metres.

(H p250)

3 *ABCD* is a rectangle, where *AD* = 5 cm and *AB* = 8 cm.

a Accurately draw the rectangle.

b Draw the angle bisector of angle *A*.

c Shade the region that satisfies the two conditions

— within 4 cm from *A*

— is closer to *AD* than *AB*.

(H p250)

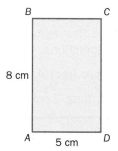

4 **a** Construct a triangle *ABC*, such that *AC* = 5.5 cm, *AB* = 10 cm and angle *B* = 30°.

b Measure angle *C*.

c Is the information you have been given enough to ensure the triangle you have drawn is unique?

(H+ p158)

5 Using compasses, Ben constructs the line PQ.

He keeps the compasses at a fixed length throughout the construction, and uses the points *A* and *B* as centres of the arcs.

a Prove that triangles *PAQ* and *PBQ* are congruent.

b Hence or otherwise, prove *M* is the midpoint of *AB*.

(H+ p158)

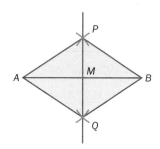

6 The diagram represents a triangular garden *ABC*.

The scale of the diagram is 1 cm represents 1 m.

A tree is planted in the garden so that it is

● nearer to *AB* than to *AC*

● within 5 m of point *A*.

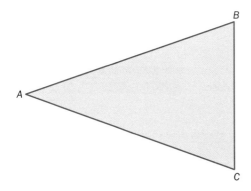

On a copy of the diagram, shade the region where the tree may be planted.

(Edexcel Ltd., 2003) 3 marks

7

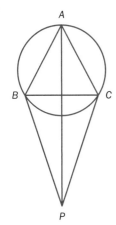

A, *B* and *C* are three points on the circumference of a circle.

Angle *ABC* = angle *ACB*.

PB and *PC* are tangents to the circle from the point *P*.

a Prove that triangle *APB* and triangle *APC* are congruent.

Angle *BPA* = 10°.

b Find the size of angle *ABC*.

(Edexcel Ltd., 2004) 7 marks

● You can use **Pythagoras' theorem** to calculate a length in a **right-angled** triangle.

$$a^2 + b^2 = c^2$$

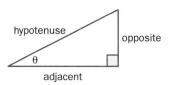

Keywords
Cosine
Pythagoras
Right-angled
Sine
Tangent

● You can use the trigonometric ratios to calculate a length or an angle in a right-angled triangle.

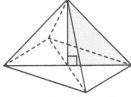

Use these when you know 2 sides and 1 angle.

$$\sin\theta = \frac{\text{opp}}{\text{hyp}} \qquad \cos\theta = \frac{\text{adj}}{\text{hyp}} \qquad \tan\theta = \frac{\text{opp}}{\text{adj}}$$

You need to remember SOHCAHTOA.

● Right-angled triangles can be drawn inside 3-D shapes.

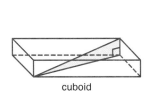

cuboid   square-based pyramid

● You can use the sine rule to calculate a length or an angle in a triangle that is NOT right-angled.

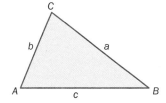

Use this when you know 2 sides and 2 angles.

$$\frac{a}{\sin A} = \frac{b}{\sin B} = \frac{c}{\sin C}$$

● You can use the cosine rule to calculate a length or an angle in a triangle that is NOT right-angled.

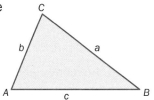

Use this when you know 3 sides and 1 angle.

$$a^2 = b^2 + c^2 - 2bc \cos A$$

● Pythagoras' theorem enables you to calculate the distance AB between two points $A(x_1, y_1)$ and $B(x_2, y_2)$.

The coordinates of the midpoint M of AB are $\left(\dfrac{x_1+x_2}{2}, \dfrac{y_1+y_2}{2}\right)$.

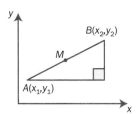

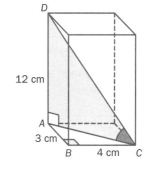

Example

A cuboid measures 3 cm by 4 cm by 12 cm.

Calculate **a** the longest diagonal CD

   b the angle ACD.

a Use Pythagoras' theorem in triangle ABC

$AC^2 = 3^2 + 4^2$

   $= 9 + 16$

   $= 25$

$AC = 5$ cm

b Use Pythagoras' theorem in triangle ACD

$CD^2 = 5^2 + 12^2$

   $= 25 + 144$

   $= 169$

$CD = 13$ cm

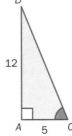

$\tan C = \dfrac{12}{5} = 2.4 \longrightarrow C = 67.4°$ (to 1 decimal place)

Example

ABC is a triangle.

$AC = 12.5$ cm.

$AB = 5.2$ cm.

$BC = 9.6$ cm.

Calculate the angle ABC.

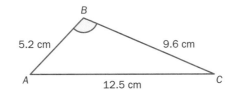

Use the cosine rule.

$12.5^2 \quad = 5.2^2 + 9.6^2 - 2 \times 5.2 \times 9.6 \times \cos B$

$156.25 \quad = 27.04 + 92.16 - 99.84 \times \cos B$

$156.25 \quad = 119.2 - 99.84 \times \cos B$

$99.84 \cos B = -37.05$

$\cos B \quad = -0.3710937$

$\quad B = 111.8°$ (to 1 decimal place)

Example

ABC is a triangle.

$AC = 8.7$ cm.

Angle $ABC = 68°$.

Angle $BAC = 65°$.

Calculate the distance BC.

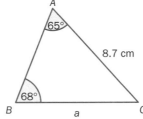

Give your answer to suitable degree of accuracy.

Use the sine rule.

$\dfrac{a}{\sin 65°} = \dfrac{8.7}{\sin 68°} \longrightarrow a = \dfrac{8.7 \times \sin 65°}{\sin 68°} = 8.5$ cm (to 1 dp)

Exercise S8

1 Calculate the distance x.

Give your answer to a suitable degree of accuracy.

(H p352)

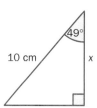

2 Harbour A is north of Lookout B.

Lighthouse C is 8.4 km east of Lookout B.

Lighthouse C is 12.5 km from Harbour A.

Calculate the bearing of

a Lighthouse C from Harbour A

b Harbour A from Lighthouse C.

(H p352)

See S1 for bearings.

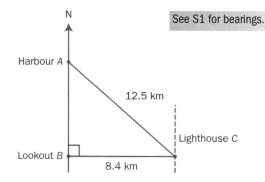

3 The point A has coordinates $(-8, 10)$.

The point B has coordinates $(10, 4)$.

Find the coordinates of the midpoint M of the line segment AB.

(H p202, H+ p194)

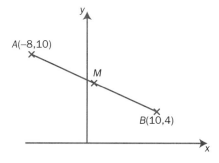

4 A square-based pyramid is shown.

The base of the pyramid is a square $BCDE$.

The length of the diagonal CE is 20 cm.

a Calculate the length of DE.

M is the midpoint of CD.
X is the centre of the square.
b Calculate the length MX.

The vertical height of the pyramid $AX = 10$ cm.
c Calculate the size of the angle that the line AM makes with the plane $BCDE$.

(H+ p200, 352)

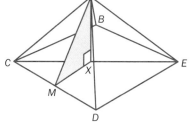

5 ABC is a triangle.

$AC = 5.6$ cm.

$AB = 12.4$ cm.

Angle $ACB = 100°$.

Calculate the angle ABC.

(H+ p244)

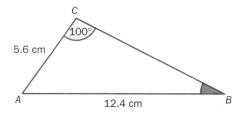

6 The area of the triangle is 8 cm².

a Calculate the value of θ.

b Calculate the perimeter of the triangle.

(H+ p246)

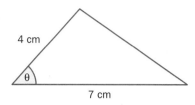

See S4 for
Area of a triangle = $\frac{1}{2}ab\sin C$.

7

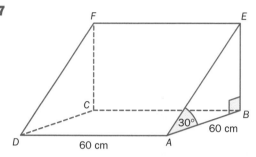

Diagram **NOT** accurately drawn

The diagram represents a prism.
AEFD is a rectangle.
ABCD is a square.

EB and FC are perpendicular to plane ABCD.
AB = 60 cm
AD = 60 cm
Angle ABE = 90°
Angle BAE = 30°

Calculate the size of the angle that the line DE makes with the plane ABCD.

Give your answer correct to 1 decimal place.

(*Edexcel Ltd., 2004*) 4 marks

8

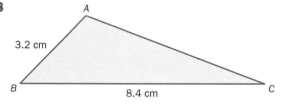

Diagram **NOT** accurately drawn

AB = 3.2 cm

BC = 8.4 cm

The area of the triangle ABC is 10 cm².

Calculate the perimeter of triangle ABC.

Give your answer correct to three significant figures.

(*Edexcel Ltd., 2004*) 6 marks

Keywords
Biased
Frequency table
Questionnaire
Random sample
Stratified sample
Survey
Systematic sample
Two-way table

● You can collect information or data through observation, controlled experiment, data logging, questionnaires and surveys.

● This **frequency table** summarises the results from a **survey**.

'Do you prefer a take-away, restaurant meal or neither?'

Meal	Frequency
Take-away	6
Restaurant	4
Neither	1

● This **questionnaire** collects two different types of data, gender and preferred meal.

You must be careful how you word your question and response section.

Questions or response sections that are confusing will not give reliable data.

Male ☐ Female ☐
Which do you prefer a
Take-away meal ☐
Restaurant meal ☐
Neither? ☐

● This **two-way table** shows the results in more detail and links two types of data, gender and preferred meal.

	Male	Female
Take-away	2	4
Restaurant	1	3
Neither	0	1

● A sample is used when you cannot collect data from all the population.

You can use different methods to choose the sample.

● In a **random sample**, each person must be equally likely to be chosen.

Picking names out of a hat is a simple way to obtain a random sample.

The larger the size of the sample, the more accurate the data will be.

● In a **stratified sample**, you divide the population into groups.

Then the number of people you select from each group is in the same proportion as the size of the group.

Finally you randomly select the required number of people from each group.

● In a **systematic sample**, you use a system to choose the sample from an ordered list.

You choose a starting position at random, then take every nth item thereafter.

n could be 10, for example.

● The sample is **biased** if each person is not equally likely to be chosen.

Choosing the first 50 students in a register is biased, as the other students would not have the opportunity to take part in the survey.

'How old are you?'

Age	Tally	Frequency
$0 < h \leqslant 20$		
$20 < h \leqslant 40$		
$40 < h \leqslant 60$		
Over 60		

● You can group numerical data into class intervals to reduce the size of the data-collection sheet.

Example

Di wants to know how often teenagers go to the cinema.

She gives her friends a questionnaire.

a Make a criticism of her method of sampling.

Part of the questionnaire is shown.

b Make two criticisms of the question and three criticisms of the response section.

> How often do you go to the cinema?
>
> 1–3 ☐
> 3–5 ☐
> 5–8 ☐

c Rewrite the question and the response section.

a The sample is biased as her friends do not represent all teenagers.

b There is no mention of a time scale in the question, e.g. a week.
The question would be better asking for recent past information.
The response section does not allow 'None'.
The response section has overlaps at 3 and 5.
The response section does not allow for 'Over 8'.

c The data is based on fact.

It gives a recent time period.

The extreme values are allowed for.

There are no overlaps.

> How often did you go to the cinema last week?
>
> None ☐
> 1–3 ☐
> 4–6 ☐
> Over 6 ☐

Example

A year group consists of Class A, Class B and Class C.

The number of students in each class is shown.

Class	Frequency
A	18
B	25
C	27

Work out the number of students needed from each class to give a stratified sample of 10 students.

Sample from Class A = $\frac{18}{70} \times 10 = 2.6 \longrightarrow$ 3 students

Sample from Class B = $\frac{25}{70} \times 10 = 3.6 \longrightarrow$ 4 students

Sample from Class C = $\frac{27}{70} \times 10 = 3.9 \longrightarrow$ 4 students

However 3 + 4 + 4 = 11 and so an adjustment must be made.

Class A   2 students

Class B   4 students

Class C   4 students

Check
2 + 4 + 4 = 10

Exercise D1

1 The times taken, in seconds, for 40 athletes to run 200 metres are shown.

43.0	31.6	35.8	33.6	25.3	34.5	36.0	30.0	32.5	26.7
31.4	38.8	30.5	36.1	27.3	35.0	31.1	21.8	37.5	34.6
25.3	39.3	24.7	30.1	40.0	30.6	44.8	39.0	31.9	28.4
34.9	29.1	33.3	38.2	21.9	32.8	36.3	28.1	38.6	34.2

a Copy and complete the frequency table.

Time (sec)	Tally	Frequency
$20 < t \leqslant 25$		
$25 < t \leqslant 30$		
$30 < t \leqslant 35$		
$35 < t \leqslant 40$		
$40 < t \leqslant 45$		

b Which class interval has the greatest number of athletes?

(H p232)

2 Kath is an aerobics instructor.

Her class consists of 2 men and 22 women, aged between 20 and 59.

One man is under 40, and 15 women are aged from 40 to 59.

a Copy and complete the two-way table to illustrate this information.

Age (years)	Men	Women	Total
$20 \leqslant a < 40$			
$40 \leqslant a < 60$			
Total			

b Calculate the percentage of the class that are under 40.

(H p66)

3 The headteacher of an 11—16 school decides one student from each form in Year 11 should give their opinion of their education over the last 5 years.

She wants to choose the students randomly.

She decides to select the first student on each form register.

a Explain why she will not obtain a random sample using this method.

b Suggest a method that will give a random sample.

c How can she improve the reliability of the information she obtains?

(H p64, H+ p62)

4 The manager of hotel wants to use a questionnaire to analyse people's use and views of hotels.

On a Monday morning, he carries out the survey in his hotel.

a Give two reasons why the survey is not representative of all people's views.

One of the questions asks, 'I'm sure you will agree this hotel is good value for money'.

b Explain why this question is unsatisfactory.

(H p62, H+ p62)

5 A machine in a paper clip factory develops a fault.
Some of the paper clips are the wrong shape.

The engineer fixes the machine and checks whether his work has been successful by using a systematic sample.

He generates a random number by throwing a dice and then looks at every 100th paper clip after the random number.

Explain why this may not be a good method to check his work.

(H+ p62)

6 Janie wants to collect information about the amount of sleep the students in her class get.

Design a suitable question she could use.

(*Edexcel Ltd., 2005*) 2 marks

7 Martin won the 400 metre race in the school sports with a time of 1 minute.

The distance was correct to the nearest centimetre.

The time was correct to the nearest tenth of a second.

a Work out the upper bound and the lower bound of Martin's speed in km/h.

See N4 for upper and lower bounds.

b Write an appropriate value for Martin's speed in km/h. Explain your answer.

The table shows the number of people in each age group who watched the school sports.

Age group	0–16	17–29	30–44	45–59	60+
Number of people	177	111	86	82	21

Martin did a survey of these people.
He used a stratified sample of 50 people according to age group.

c Work out the number of people from each age group that should have been in his sample of 50.

Copy and complete the table.

Age group	0–16	17–29	30–44	45–59	60+	Total
Number of people in sample						

(*Edexcel Ltd., 2004*) 9 marks

● **Discrete data** can only take exact values.

   The number of people in a car is an example of discrete data.

● You can use a horizontal or vertical **bar chart** to display data.

   The bars have equal thickness and are separated by gaps.

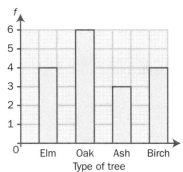

The mode is Oak – see D4.

● A **pie chart** gives a visual display of all the data.

   The size of the sector angle is proportional to the frequency.

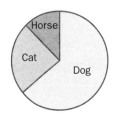

The mode is Dog – see D4.

● You can use a **stem-and-leaf diagram** to display numerical data.

   A stem-and-leaf diagram is similar to a horizontal bar chart, but with more detail.

   The data is in numerical order for an **ordered** stem-and-leaf diagram.

A key is essential.

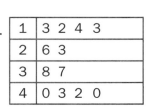

Key: | 1 | 3 | means 13

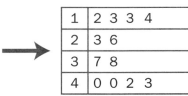

Key: | 1 | 3 | means 13

● A **scatter graph** is used to test if there is a linear relationship between two sets of data.

   The data is collected in pairs and plotted as coordinates.

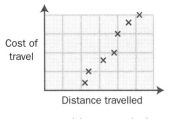

positive correlation

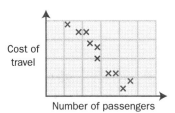

negative correlation

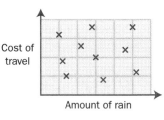

no correlation

There is a linear relationship between Cost of travel and Distance travelled.

There is a linear relationship between Cost of travel and Number of passengers.

There is no linear relationship between Cost of travel and Rainfall.

● You can draw a **line of best fit** if there is a strong linear relationship.

   The line can be used to predict data values within the range of data collected.

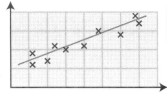

The line of best fit does not have to pass through (0,0).

Fifty photographers were asked to name their favourite season for taking photographs.

The results are shown in the frequency table.

a Draw a pie chart to illustrate the data.

b State the modal season.

Season	Frequency
Spring	11
Summer	7
Autumn	24
Winter	8

a Spring $\frac{11}{50} \times 360° = 79.2° \rightarrow 79°$

   Summer $\frac{7}{50} \times 360° = 50.4° \rightarrow 50°$

   Autumn $\frac{24}{50} \times 360° = 172.8° \rightarrow 173°$

   Winter $\frac{8}{50} \times 360° = 57.6° \rightarrow 58°$

   Check: $79° + 50° + 173° + 58° = 360°$

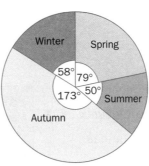

b Autumn (the most frequent choice of season)

Rick owns an ice cream van. He sells ice creams and cups of tea.

The number of ice creams and cups of tea he sells each day during one week are shown in the frequency table.

Number of ice creams	100	20	30	10	80	60	50
Number of cups of tea	20	70	60	72	30	46	48

a Draw a scatter graph to show the data.

b State the type of correlation.

c Draw the line of best fit.

d If Rick sells 40 ice creams, how many cups of tea should he expect to sell?

e Why would it not be sensible to use the line of best fit to estimate the cups of tea sold when 120 ice creams are sold?

a, c

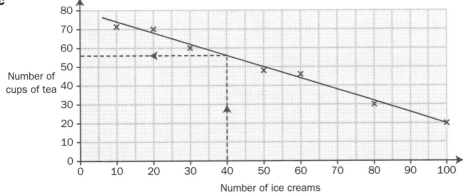

b Negative correlation

d Using the line of best fit, 56 cups of tea

e 120 cups of tea is outside the range of the given data and so the line of best fit is unreliable at 120 cups of tea.

Exercise D2

1 The scores of 20 ice skaters in a competition are shown.

4.8   5.0   5.8   5.7   5.6   5.7   4.9   5.0   6.0   5.3

5.1   5.2   5.4   4.7   4.5   5.8   5.7   5.8   5.9   6.0

Copy and complete an ordered stem-and-leaf diagram to show the data.

Remember to give the key.

2 Four foods are tested for fat content.

The results are shown in the frequency table.

Draw a pie chart to illustrate the data.

Show your working.

Food	Fat (g) per 100 g
Mayonnaise	80
Nuts	76
Peanut butter	50
Crisps	34

3 Eight people are tested for their speed of texting on a mobile phone.

Their average speed is calculated measured in characters per minute.

They are then each given the same standard message to text and the time taken is recorded.

The results are shown in the frequency table.

Speed (characters per min)	33	45	20	35	10	49	40	25
Time taken (seconds)	30	20	35	26	36	23	27	30

a Draw a scatter diagram to show the data.

b State the type of correlation.

Scatter diagram and scatter graph mean the same.

c Draw the line of best fit.

d If Sarah's speed of texting is 30 characters per minute, how long would you expect Sarah to take to text the standard message?

Robin's speed of texting is 70 characters per minute.

e You could use the line of best fit to estimate the time Robin should take to text the standard message.

Why is this not a sensible method?

(H p112, H+ p238)

4 The number of mountains over 3000 feet in each country are shown in the frequency table.

Draw a pie chart to illustrate the data.

Show your working.

Country	Frequency
England	4
Scotland	284
Wales	8
Ireland	7

5 Gavin works behind a bar in Spain.

The daily temperature and the number of cold drinks he sells in one week are shown in the frequency table.

Temperature (°C)	20	29	15	25	12	30	35
Number of cold drinks	125	225	80	175	45	185	250

a Draw a scatter graph to show the data.

b State the type of correlation.

c Draw the line of best fit.

d If the temperature is 24 °C, how many cold drinks should Gavin expect to sell?

e Why would using the line of best fit to estimate the number of cold drinks sold for a temperature of 40 °C not be sensible?

(H p112, H+ p238)

6 Here are the times, in minutes, taken to change some tyres.

5	10	15	12	8	7	20	35	24	15
20	33	15	25	10	8	10	20	16	10

Draw a stem-and-leaf diagram to show these times.

(*Edexcel Ltd., 2003*) 3 marks

7 The table gives information about the medals won by Austria in the 2002 Winter Olympic games.

Medal	Frequency
Gold	3
Silver	4
Bronze	11

Draw an accurate pie chart to show this information.

(*Edexcel Ltd., 2005*) 4 marks

Keywords
Continuous data
Cumulative frequency
Frequency density
Frequency polygon
Histogram
Median
Quartiles

● **Continuous data** can take any value and cannot be measured exactly.

The weight of a person is an example of continuous data.

● You can use a **histogram** to display grouped continuous data.

This histogram has equal class intervals.

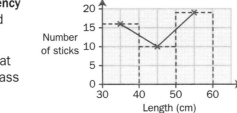

No gaps between the bars.

● You can also use a **frequency polygon** to display grouped continuous data.

The points are plotted at the midpoints of the class intervals.

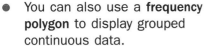

Just draw the polygon, not the dotted lines.

● This histogram does not have equal class intervals.

The heights of the bars have been adjusted to allow for the different class widths.

Frequency density = $\dfrac{\text{frequency}}{\text{class width}}$

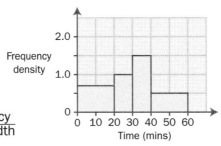

The vertical axis is Frequency density.

● You can represent large sets of grouped data on a **cumulative frequency** diagram.

Weight (kg)	Cumulative frequency
$0 < w \leqslant 5$	8
$5 < w \leqslant 10$	22
$10 < w \leqslant 15$	55
$15 < w \leqslant 20$	70
$20 < w \leqslant 25$	75
$25 < w \leqslant 30$	78
$30 < w \leqslant 35$	80

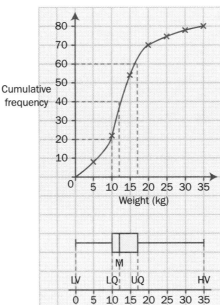

Points are plotted at the upper value of each class interval.

The median, lower quartile and upper quartile can be found from the cumulative frequency diagram.

The **box plot** can be drawn using the cumulative frequency diagram.

See D4 for median, quartiles and box plot.

Example

The Great North Run is a 13-mile race.

The times taken for 80 runners to complete the race are shown.

Time (mins)	Frequency
$50 < t \leqslant 60$	5
$60 < t \leqslant 70$	9
$70 < t \leqslant 80$	38
$80 < t \leqslant 90$	20
$90 < t \leqslant 100$	8

a Copy and complete the cumulative frequency table.

b Draw a cumulative frequency diagram for this data.

Time (mins)	Cumulative frequency
$\leqslant 60$	5
$\leqslant 70$	
$\leqslant 80$	
$\leqslant 90$	
$\leqslant 100$	

c Calculate the median and interquartile range.

d Draw the box plot for this data.

e Estimate the number of runners who took longer than 88 minutes to run the race.

---

a

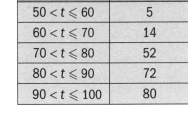

Time (mins)	Cumulative frequency
$50 < t \leqslant 60$	5
$60 < t \leqslant 70$	14
$70 < t \leqslant 80$	52
$80 < t \leqslant 90$	72
$90 < t \leqslant 100$	80

b

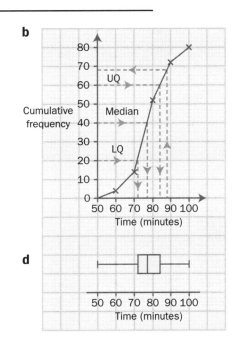

c From the graph, Median = 77 mins

Upper quartile = 84 mins

Lower quartile = 72 mins

Interquartile range = 84 − 72

= 12 mins

d

e From the graph, 68 runners ran in 88 mins or less.

Number of runners taking over 88 mins = 80 − 68

= 12 runners

Exercise D3

1 100 people take part in a survey about the number of hours a week their television is switched on. None of the people watched the TV for 60 hours or more.

The results of the survey are shown in this very WRONG graph.

This graph is very **WRONG**.
Don't be tempted to draw one like this!

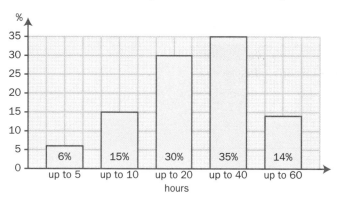

a Make three criticisms of the graph.

b Copy and complete the table and draw a correct histogram to show the data.

Time (hours)	$0 \leqslant t < 5$	$5 \leqslant t < 10$	$10 \leqslant t < 20$	$20 \leqslant t < 40$	$40 \leqslant t < 60$
Class width					
Frequency	6	15	30	35	14
Frequency density					

(H+ p266—270)

2 The histogram gives information about the times, in minutes, 135 students spent on the Internet last night.

Use the histogram to copy and complete the table.

Time (*t* minutes)	Frequency
$0 < t \leqslant 10$	
$10 < t \leqslant 15$	
$15 < t \leqslant 30$	
$30 < t \leqslant 50$	
TOTAL	**135**

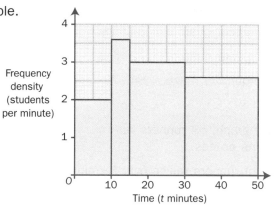

(*Edexcel Ltd., 2004*) 2 marks

3 The table shows information about the number of hours that 120 children used a computer last week.

Number of hours (h)	Frequency
$0 < h \leqslant 2$	10
$2 < h \leqslant 4$	15
$4 < h \leqslant 6$	30
$6 < h \leqslant 8$	35
$8 < h \leqslant 10$	25
$10 < h \leqslant 12$	5

a Copy and complete the cumulative frequency table.

Number of hours (h)	Cumulative frequency
$0 < h \leqslant 2$	10
$2 < h \leqslant 4$	
$4 < h \leqslant 6$	
$6 < h \leqslant 8$	
$8 < h \leqslant 10$	
$10 < h \leqslant 12$	

b On a copy of the grid, draw a cumulative frequency graph for your table.

> Remember to plot the points at the upper values of each class interval.

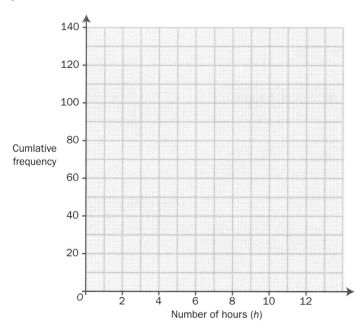

c Use your graph to find an estimate for the number of children who used a computer for **less** than 7 hours last week.

(Edexcel Ltd., 2005) 5 marks

Keywords
Average
Box plot
Estimate of the mean
Interquartile range
Modal class
Moving average
Quartiles
Spread

- You can represent a set of numerical data using

  a measure of **spread**   —the range

                           —the **interquartile range**

  a measure of **average**   —the mean

                           —the mode or modal class

                           —the median.

- You can calculate the mean, mode and median for a **frequency table**.

Number	Frequency
1	10
2	20
3	36
4	34

See the example.

- You cannot calculate the exact mean for grouped data.

  You can calculate an **estimate of the mean**.

Height	Frequency
$0 < h \leqslant 100$	14
$100 < h \leqslant 200$	27
$200 < h \leqslant 300$	9

Use the mid-values of each class interval.

- You cannot calculate the mode for grouped data.

  You can state the **modal class**.

Weight	Frequency
$0 < w \leqslant 10$	4
$10 < w \leqslant 20$	8
$20 < w \leqslant 30$	9

The modal class is $20 < w \leqslant 30$ as this class has the highest frequency.

- You cannot calculate the median for grouped data.

  You can find the class in which the median lies.

Number	Frequency
51–55	4
56–60	8
61–65	9

The median is the 11th number out of 21.
This number is in the 56–60 class interval.

- A **box plot** gives a visual display of a set of data.

  The median, upper **quartile** and lower quartile, and the highest and lowest values are shown.

See D3 for a box plot for a large set of data.

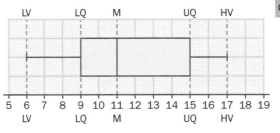

- You can calculate a **moving average** for time series data.

  The moving averages smooth out the fluctuations of the data.

See D5 for time series graphs and moving averages graphs.

2006				2007			
March	**June**	**Sept**	**Dec**	**March**	**June**	**Sept**	**Dec**
3	6	7	2	4	5	8	1

The first 4-point moving average is the mean of 3, 6, 7 and 2.

The second 4-point moving average is the mean of 6, 7, 2 and 4.

Example

Eleven students are given a spelling test of 20 words.

The scores for the test are

   9   10   12   13   14   14   15   17   18   20   20

a Calculate   **i** the median

                ii the range

                iii the upper and lower quartiles

                iv the interquartile range.

b Draw a box plot to show the information.

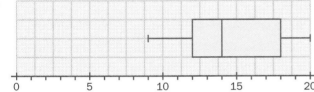

a **i** Median = 14 $\left(\text{middle value when arranged in order or}\right.$

   $\frac{1}{2}$(11+1)th term$\Big)$

  ii Range = 20 − 9 = 11 (highest value minus lowest value)

  iii Upper quartile = 18 $\left(\frac{3}{4}(11+1)\text{th term}\right)$

     Lower quartile = 12 $\left(\frac{1}{4}(11+1)\text{th term}\right)$

  iv Interquartile range = 18 − 12 = 6 (UQ minus LQ)

The numbers are already in order.

b

Leroy goes ten-pin bowling. He rolls the ball 50 times.

The frequency table shows the number of pins he knocked down each roll.

Number of pins	0	1	2	3	4	5	6	7	8	9	10
Frequency	0	3	2	1	8	12	10	14	0	0	0

a Calculate the total number of pins knocked down.

b Calculate the mean number of pins knocked down.

c State the mode.

d Calculate the range.

a (0 × 0) + (1 × 3) + (2 × 2) + (3 × 1) + (4 × 8) + (5 × 12) + (6 × 10) + (7 × 14) + (8 × 0) + (9 × 0) + (10 × 0) = 260

b Mean = 260 ÷ 50 = 5.2 pins

c Mode = 7 pins (the most frequent value)

d Range = 7 − 1 = 6 pins (highest value minus lowest value)

Exercise D4

1 The stem-and-leaf diagram shows the number of lessons needed before 15 people passed their driving test.

0	4 6 9
1	3 5 8 8 8 9
2	2 2 4
3	2 2 3

Key: | 1 | 3 | means 13 people

a Calculate the mean number of driving lessons.

b Find the median and upper and lower quartiles.

c Calculate the range and the interquartile range.

d State the mode.

e Draw a box plot of this data.

(H p118, H+ p110)

2 The number of bedrooms in each house on a street are shown in the frequency table.

Number of bedrooms	Number of houses
1	3
2	8
3	4
4	5

a Calculate **i** the total number of bedrooms in the street

ii the total number of houses on the street

iii the mean number of bedrooms per house.

b State the mode.

(H p230, H+ p66)

3 A survey of 50 shoppers records the value of the shopping in their baskets.

The results are shown in the frequency chart.

a State the modal class.

b Find the class interval that contains the median.

c Calculate an estimate of the mean.

(H p232, H+ p70)

Value of the shopping	Number of customers
£0 < v ⩽ £5	3
£5 < v ⩽ £10	12
£10 < v ⩽ £15	15
£15 < v ⩽ £20	16
£20 < v ⩽ £25	4

4 The Number 86 bus has 16 passengers and the Number 42 bus has 9 passengers.

Each passenger is asked how long he or she had to wait for the bus.

The mean waiting time for the Number 86 bus passengers was 3 minutes.

The mean waiting time for the Number 42 bus passengers was 5 minutes.

Calculate the mean waiting time for all 25 passengers.

(H p70, H+ p68)

5 The table shows the money Jane spends each month on petrol for her car.

Jan	Feb	Mar	Apr	May	Jun	Jul	Aug	Sep	Oct	Nov	Dec
£26	£42	£30	£50	£58	£54	£42	£46	£54	£32	£26	£18

a Calculate the nine 4-month moving averages for this information.

b Which four months have the lowest moving average?

(H p238, H+ p234)

6 Mary recorded the heights, in centimetres, of the girls in her class.

She put the heights in order.

132   144   150   152   160   162   162   167

167   170   172   177   181   182   182

a Find **i** the lower quartile

ii the upper quartile.

b On the grid, draw a box plot for this data.

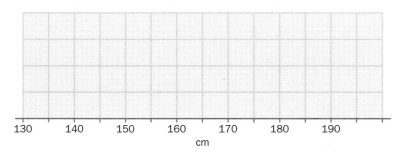

(*Edexcel Ltd., 2003*) 2 marks

7 The table shows the number of computer games sold in a supermarket each month from January to June.

Jan	Feb	Mar	Apr	May	Jun
147	161	238	135	167	250

a Work out the 3-month moving averages for this information.

In a sale, a supermarket took 20% off its normal prices.

On Fun Friday, it took 30% off its sale prices.

Fred says, 'That means there was 50% off the normal prices'.

b Fred is wrong. Explain why.

See N7 for percentage calculations.

(*Edexcel Ltd., 2004*) 4 marks

- You can compare two sets of numerical data using

    a measure of **spread** — the range
    — the interquartile range

    a measure of **average** — the mean
    — the mode or modal class
    — the median.

Keywords
Average
Box plot
Frequency polygon
Line graph
Moving-averages graph
Spread
Trend

- You can compare two sets of numerical data using **box plots.**

    Two box plots of related data can be shown on the same graph paper.

    Data A has a larger range.

    Data B has a lower median.

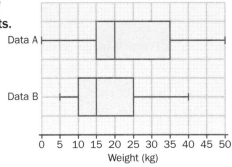

See D4 for range, interquartile range, mean, mode, modal class and median.

See D4 for box plots.

- You can compare two sets of data using graphs.

    Two **frequency polygons** of related data can be shown on the same graph.

    Data A has a larger range.

    Data B has a lower modal class.

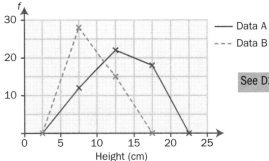

See D2 and D3 for frequency polygons.

- You use a **line graph** to show how data change with time.

    The number of stolen cars is shown for each successive year.

    You can see the **trend** from the graph.

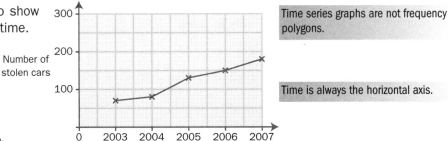

Time series graphs are not frequency polygons.

Time is always the horizontal axis.

- You can draw a **moving-averages graph** for time series data.

    The moving-averages graph smoothes out the fluctuations of the raw data.

    The moving averages are plotted on the same axes as the original data, at the midpoint of the time span of the moving average.

    By extending the moving-averages graph, you can predict the trend.

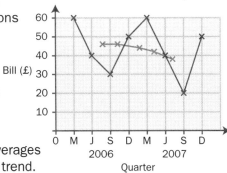

See D4 for moving averages.

Example

Ten people are asked to shuffle a pack of cards with their right hand and then their left hand. The times taken, in seconds, are shown in the stem-and-leaf diagram.

Right hand		Left hand
8 5	0	
6 5 4 4	1	8
8 2 1	2	3 6 7 7 7
0	3	2 4 9 9

a Calculate the mean and range for each set of data.

b Make **two** comparisons of the right hand and left hand times. Key: 4/1/8 means 14 seconds Right 18 seconds left

a Right hand Mean = (8 + 5 + 14 + 14 + 15 + 16 + 21 + 22 + 28 + 30) ÷ 10

= 173 ÷ 10 = 17.3 seconds

Left hand Mean = (18 + 23 + 26 + 27 + 27 + 27 + 32 + 34 + 39 + 39) ÷ 10

= 292 ÷ 10 = 29.2 seconds

Right hand Range = 30 − 5 = 25 seconds

Left hand Range = 39 − 18 = 21 seconds

b On average, the left hand took longer to shuffle the cards (using the mean).

The right hand times are more spread out than the left hand times.

Example

The year-on-year percentage of unemployed people in the UK is shown.

Year	2000	2001	2002	2003	2004	2005	2006	2007
%	5.5	5.1	5.2	5.0	4.8	4.7	4.0	?

a Draw a line graph to show this information.

b Calculate the 3-year moving averages from this data.

c Plot the 3-year moving averages on graph paper.

d Estimate a value for the 3-year moving average to be plotted at 2006.

e Estimate the percentage of unemployed people in 2007.

a, c

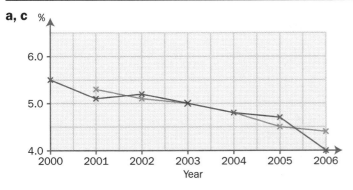

b 2001 (5.5 + 5.1 + 5.2) ÷ 3 = 15.8 ÷ 3 = 5.3

2002 (5.1 + 5.2 + 5.0) ÷ 3 = 15.3 ÷ 3 = 5.1

2003 (5.2 + 5.0 + 4.8) ÷ 3 = 15.0 ÷ 3 = 5.0

2004 (5.0 + 4.8 + 4.7) ÷ 3 = 14.5 ÷ 3 = 4.8

2005 (4.8 + 4.7 + 4.0) ÷ 3 = 13.5 ÷ 3 = 4.5

d About 4.4 (found by extending the moving averages graph to 2006)

e (4.7 + 4.0 + x) ÷ 3 = 4.4 gives x = 4.5%

115

Exercise D5

1   The frequency polygons show the weights of some Year 11 students.

See D3 for frequency polygons.

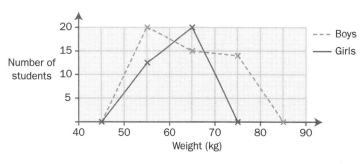

Make **two** comparisons between the two distributions.

(H p234, H+ p232)

2   Alan and Zena want to compare the house prices where each one lives.

The price of 100 houses in Alan's and Zena's neighbourhood are shown in the frequency tables.

Alan

Price in £1000s	Frequency
$0 < p \leqslant 50$	15
$50 < p \leqslant 100$	25
$100 < p \leqslant 150$	38
$150 < p \leqslant 200$	18
$200 < p \leqslant 250$	4
$250 < p \leqslant 300$	0

Zena

Price in £1000s	Frequency
$0 < p \leqslant 50$	10
$50 < p \leqslant 100$	14
$100 < p \leqslant 150$	16
$150 < p \leqslant 200$	24
$200 < p \leqslant 250$	26
$250 < p \leqslant 300$	10

a   Copy and complete the cumulative frequency tables for each set of data.

Alan

Price in £1000s	Cumulative frequency
$\leqslant 50$	15
$\leqslant 100$	40
$\leqslant 150$	
$\leqslant 200$	
$\leqslant 250$	
$\leqslant 300$	

Zena

Price in £1000s	Cumulative frequency
$\leqslant 50$	10
$\leqslant 100$	
$\leqslant 150$	
$\leqslant 200$	
$\leqslant 250$	
$\leqslant 300$	

See D3 for cumulative frequency.

b   Draw cumulative frequency diagrams for each set of data.

c   Calculate the median and interquartile range for each distribution.

d   Make **two** comparisons between the house prices in Alan's and Zena's neighbourhood.

(H p266—270, H+ p112, 114, 118)

3 The cost of hiring a cottage has risen year on year.

The weekly cost is shown in the frequency table.

Year	2000	2001	2002	2003	2004	2005	2006	2007
Cost (£)	240	272	300	340	352	380	400	?

a Draw a line graph to show this information.

b Calculate the 4-year moving averages from this data.

c Plot the 4-year moving averages on graph paper.

d Estimate a value for the 4-year moving average to be plotted at 2005/6.

e Estimate the cost of hiring a cottage in 2007.

(H p238, H+ p236)

4 Gareth pays his electricity and gas bill separately every month.

The box plots summarise the monthly payments during one year. See D4 for box plots.

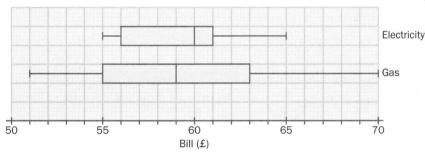

a Use the box plots to calculate

  i the ranges

  ii the interquartile ranges

  iii the medians.

b Write **two** statements to compare Gareth's electricity and gas bills.

(H p274, H+ p118)

5 The heights of two types of rose plants, Miniature and Standard, are summarised in these histograms. See D3 for histograms.

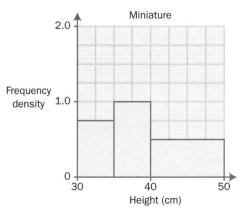
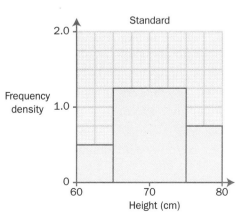

Use the histograms to make **two** comparisons between the two distributions.

(H+ p272)

D6 Probability

Keywords
Biased
Equally likely
Event
Expected frequency
Independent
Mutually exclusive
Outcome
Probability tree diagram
Relative frequency
Sample space diagram

● A trial is an activity.

Picking one ball from a bag of red and white balls is a trial.

● An **outcome** is one possible result of a trial.

All the outcomes when picking the ball are red, red, red, white, white.

● Each outcome is **equally likely** as the balls are identical in size and shape.

If the balls were not the same size and shape, they would be **biased**.

These outcomes are **mutually exclusive** because you cannot pick a red and a white ball at the same time.

● An **event** is one or more outcomes of the trial.

A red ball is an event when picking a ball from the bag.

● Probability measures the chance of an event happening.

All probabilities have a value between 0 and 1.

0 means impossible.
1 means certain.

$$\text{Probability of an event happening} = \frac{\text{number of favourable outcomes}}{\text{total number of all possible outcomes}}$$

● The probabilities of mutually exclusive outcomes add up to 1.

For a dice $P(4) = \frac{1}{6}$

Probability of an event **not** happening = 1 − probability of the event happening

For two mutually exclusive events A and B,

$$P(A \text{ or } B) = P(A) + P(B)$$

For a dice $P(\text{not }4) = 1 - P(4)$

'Or' = +

● The **expected frequency** is the number of times you expect the event to happen.

Expected frequency = probability × number of trials

The estimated probability is more reliable the greater the number of trials.

● You can estimate the probability from experiments.

The estimated or experimental probability is called the **relative frequency**.

$$\text{Relative frequency} = \frac{\text{number of successful trials}}{\text{total number of trials}}$$

● You can use a **sample space diagram** or a **probability tree diagram** to show the outcomes of two successive events.

● Two or more successive events are **independent**, if when one event occurs, then it has no effect on the other event(s) occurring.

Picking successive balls from a bag and replacing the ball each time will give independent events.

For two independent events A and B, $P(A \text{ and } B) = P(A) \times P(B)$

'And' = ×

Jane spins a coin 10 times. She spins a Head 8 times. She says, 'The coin is biased.'

a Explain why she may be wrong.

Jane carries on spinning the coin.

She eventually gets 483 Heads and 517 Tails.

b Calculate the relative frequency for **i** Heads

                                                   ii Tails.

Heads   Tails

a The sample is too small.

b **i** $\dfrac{483}{1000}$ **ii** $\dfrac{517}{1000}$

$483 + 517 = 1000$

Four red cards numbered 1, 2, 3 and 4 and three white cards numbered 1, 2 and 3 are placed in separate piles.

One card is randomly selected from each pile.

a Draw a **sample space diagram** to show all the possible outcomes.

b Calculate the probability that the cards show the same number.

The cards are replaced in the correct piles.

Two cards are selected in this way 200 times.

c How many times would you expect the two cards to show the same number?

1   2   3   4

1   2   3

a

		Red card			
		1	**2**	**3**	**4**
White card	**1**	(1,1)	(1,2)	(1,3)	(1,4)
	2	(2,1)	(2,2)	(2,3)	(2,4)
	3	(3,1)	(3,2)	(3,3)	(3,4)

b $\dfrac{3}{12} = \dfrac{1}{4}$

c $\dfrac{1}{4} \times 200 = 50$ times

A bag contains 7 red balls and 3 white balls. One ball is taken out at random and **not** replaced. Another ball is then taken out.

a Draw a **probability tree diagram** to show the outcomes. Label the branches of the diagram.

b Calculate the probability that at least one ball is red.

a

1st ball	2nd ball	Outcome	Probability
	$\frac{6}{9}$ —— Red	R and R	$\frac{7}{10} \times \frac{6}{9}$
$\frac{7}{10}$ Red	$\frac{3}{9}$ —— White	R and W	$\frac{7}{10} \times \frac{3}{9}$
$\frac{3}{10}$ White	$\frac{7}{9}$ —— Red	W and R	$\frac{3}{10} \times \frac{7}{9}$
	$\frac{2}{9}$ —— White	W and W	$\frac{3}{10} \times \frac{2}{9}$

b P(At least one ball is red) $= 1 - \left(\dfrac{3}{10} \times \dfrac{2}{9}\right)$

                                   $= 1 - \dfrac{1}{15}$

                                   $= \dfrac{14}{15}$ This is very likely!

$\left(\dfrac{7}{10} \times \dfrac{6}{9}\right) + \left(\dfrac{7}{10} \times \dfrac{3}{9}\right) +$

$\left(\dfrac{3}{10} \times \dfrac{7}{9}\right) = \dfrac{42}{90} + \dfrac{21}{90} + \dfrac{21}{90}$

$= \dfrac{84}{90} = \dfrac{14}{15}$

Exercise D6

1 On Kevin's street, people either live in a flat, a town house or a semi-detached house.

a Copy and complete the two-way table that shows this information.

	Flat	Town house	Semi-detached	Total
Male	19			50
Female		9	14	
Total			30	80

b Calculate the probability that a person chosen at random from the whole street lives in a flat.

c Calculate the probability that a person chosen at random from the town house residents is male.

(H p134, H+ p134)

2 Rob rolls a dice 50 times. He records the number of sixes after 10, 20, 30, 40 and 50 rolls of the dice.

The frequency table shows the results.

Rolls of the dice	Total number of sixes	Relative frequency
After 10	2	
After 20	3	
After 30	5	
After 40	6	
After 50	8	

a Copy and complete the table to show the relative frequency after 10, 20, 30, 40 and 50 rolls of the dice.

b On graph paper, draw a line graph to show the relative frequencies.

c State, as a decimal, the theoretical probability of throwing a 6 on a dice. Mark this value on your graph.

(H p142, H+ p138)

3 On her way home, Isabella drives past two traffic lights that work independently from each other.

The traffic lights are either on Red or Green.

The probability that the first traffic light is on Red is 0.2.

The probability that the second traffic light is on Red is 0.15.

a Draw a probability tree diagram to show this information.

b Calculate the probability that both traffic lights are on Green.

(H p330, 332, 334, H+ p326, 328)

4 Here is a 4-sided spinner.

The sides of the spinner are labelled 1, 2, 3 and 4.

The spinner is biased.

The probability that the spinner will land on each of the numbers 2 and 3 is given in the table.

The probability that the spinner will land on 1 is equal to the probability that it will land on 4.

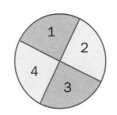

Number	1	2	3	4
Probability	x	0.3	0.3	x

a Work out the value of x.

Sarah is going to spin the spinner 200 times.

b Work out an estimate for the number of times it will land on 2.

(*Edexcel Ltd., 2005*) 4 marks

5 Jeremy designs a game for a school fair.
He has two 5-sided spinners.
The spinners are equally likely to land on each of their sides.
One spinner has 2 red sides, 1 green side and 2 blue sides.
The other spinner has 3 red sides, 1 yellow side and 1 blue side.

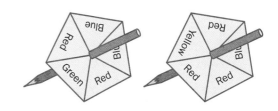

a Calculate the probability that the two spinners will land on the same colour.

The game consists of spinning each spinner once.
It costs 20p to play the game.
To win a prize both spinners must land on the same colour.
The prize for a win is 50p.
100 people play the game.

b Work out an estimate of the profit that Jeremy should expect to make.

(*Edexcel Ltd., 2005*) 5 marks

6 A bag contains 3 black beads, 5 red beads and 2 green beads.

Gianna takes a bead at random from the bag, records its colour and replaces it.

She does this two more times.

Work out the probability that, of the three beads Gianna takes, exactly two are the same colour.

(*Edexcel Ltd., 2003*) 5 marks

1 This table gives the temperatures in Leeds at different times on 12 February.

Midnight	−4℃
4 am	−6℃
8 am	−1℃
Noon	4℃
3 pm	6℃
7 pm	3℃

  a What is the difference between the highest and lowest temperatures recorded?   (*2 marks*)

  b Between 7 pm and 11 pm, the temperature dropped by 4℃. What was the temperature at 11 pm?   (*1 mark*)

2 **a** **i** Write the ratio $20:28:32$ in its simplest form.   (*1 mark*)

    ii Divide £400 in the ratio $20:28:32$.   (*1 mark*)

  b A jar contains 250g of chocolate and 150g of sugar.

    i Give the ratio of chocolate to sugar in its simplest form.   (*1 mark*)

    ii What fraction of the mixture is sugar?   (*1 mark*)

3 Simplify these as far as possible, giving your answers in index form.

  a $5^5 \times 5^3$ (*2 marks*)

  b $\sqrt{3^4 \times 5^2}$ (*2 marks*)

4 A square has side 3 cm, correct to the nearest cm. Find the lower bound for its area.   (*2 marks*)

5 The diagram shows a rectangle of area A cm^2.

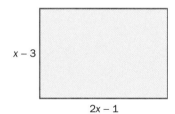

$x - 3$

$2x - 1$

  a Show that $A = 2x^2 - 7x + 3$ (*2 marks*)

  b The perimeter is 28 cm.
Find the value of A.   (*3 marks*)

6 The number of dots in these patterns form a linear sequence.

Pattern 1   Pattern 2   Pattern 3

  a Write the number of dots in pattern 4.   (*1 mark*)

  b Find an expression, in terms of n, for the nth term in the sequence.   (*2 marks*)

  c Does any pattern in the sequence contain 247 dots? You must explain your answer.   (*1 mark*)

7 **a** Complete the table of values for $y = 5 - 2x$. (*2 marks*)

x	−3	−2	−1	0	1	2	3
y	11		7				−1

b On this grid, draw the graph of $y = 5 - 2x$. (*2 marks*)

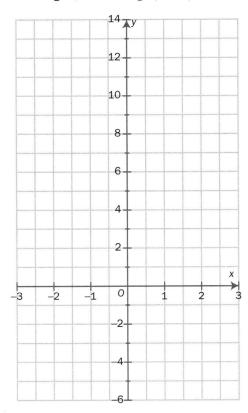

8 Here are the equations of five straight lines.

A: $y = 5x + 2$ B: $y + 3x = 1$ C: $2y - 3x = 4$

D: $y = \frac{1}{4}x$ E: $y = 4x - 3$

a The point P lies on line A. The x coordinate of P is 3. Find the y coordinate. (*1 mark*)

b Which of the lines has a negative gradient? (*1 mark*)

c Write the gradient of the line A. (*1 mark*)

d Write the equation of a different line which is parallel to line E: $y = 4x - 3$. (*1 mark*)

e Write the coordinates of the point where line B crosses the y axis. (*1 mark*)

9

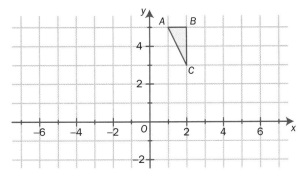

a Rotate triangle ABC 90° anticlockwise about the point (1,2).
Label the triangle DEF. (*3 marks*)

b Describe the transformation that will map triangle DEF onto ABC. (*1 mark*)

10 The angles at A, B and C are each 90°.
Work out the area of the shape.

(3 marks)

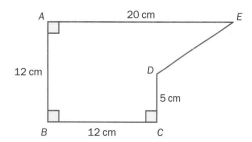

11 A school is planning to offer Key Stage 3 resources online. It wants to find out about student access to the internet outside school. It plans to do a stratified sample of 50 students by year group and gender. The numbers of boys and girls at Key Stage 3 is given in the table.

	Year 7	Year 8	Year 9
Boys	75	84	68
Girls	80	71	74

a How many Year 9 students will be in the sample? *(2 marks)*

b Explain why this is not a random sample. *(2 marks)*

12 The ages of members of an angling club are shown below.

49	35	12	62	23	34	52	11	51	57	31
48	56	67	27	45	53	51	61	63	16	

a Draw a stem-and-leaf diagram to represent this information. *(2 marks)*

b Work out the median. *(1 mark)*

c Draw a box and whisker plot for this data. Work out the median. *(4 marks)*

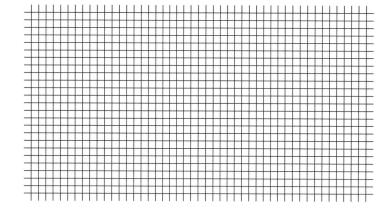

13 A bag has 3 red and 7 blue balls in it.
A second bag has 2 red and 3 blue balls in it. One ball is chosen at random
from each.

   a Complete the tree diagram. *(2 marks)*

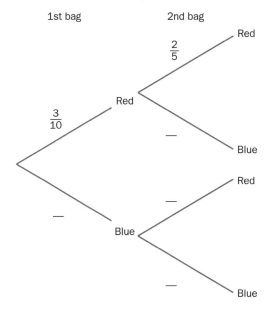

1st bag   2nd bag

   b What is the probability that at least one red ball is chosen? *(2 marks)*

14 This scatter graph gives information about the number of people vaccinated
against and infected by a certain disease.

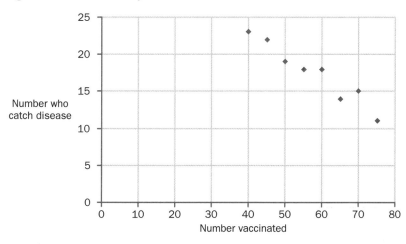

Number who catch disease

Number vaccinated

   a Describe the correlation between the number of people vaccinated in a group
and the number who catch the disease. *(1 mark)*

   b Draw a line of best fit on the diagram. *(1 mark)*

   c Another group of the same size had 52 people vaccinated.
Use your line of best fit to estimate the number of people who catch
the disease in this group. *(1 mark)*

15 The average weight of the eight forwards on a rugby team is 105.6 kg.
The average weight of the seven backs on the same team is 92.7 kg.
Calculate the average weight of the whole team. *(3 marks)*

16 **a** Does the point $P(3, 7)$ lie on the line $y = 2x - 1$? *(1 mark)*

  b The point Q lies on the line $y = 2x - 1$. The y coordinate of Q is 19.
  Find the x coordinate of Q. *(2 marks)*

  c Write the equation of a different straight line that is parallel to $y = 2x - 1$. *(2 marks)*

17 Solve these simultaneous equations.

  $7x + 4y = 17$

  $4x - 2y = 14$ *(3 marks)*

18 Expand and simplify these.

  a $7(2x + 3) - 3(x - 4)$ *(2 marks)*

  b $(2x + y)(x + 3y)$ *(2 marks)*

19 Factorise these completely.

  a $10x^2 - 15xy$ *(2 marks)*

  b $p^2 - 49q^2$ *(2 marks)*

  c $3x^2 + 7x - 6$ *(2 marks)*

20 Solve, giving your answers to 4 significant figures where appropriate.

  a $x^2 - 5x - 6 = 0$ *(3 marks)*

  b $2x^2 - 3x - 4 = 0$ *(3 marks)*

21 *DE* is parallel to *BC*.

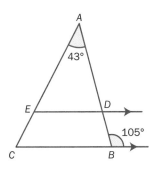

  a Work out angle *ADE*. Give a reason for your answer. *(2 marks)*

  b Work out angle *AED*. Give a reason for your answer. *(2 marks)*

22 The exterior angle of a regular polygon is 45°.

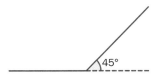

  Work out the number of sides in the regular polygon. *(2 marks)*

23 *AB* is parallel to *CD* and *X* is the midpoint of *AC*.

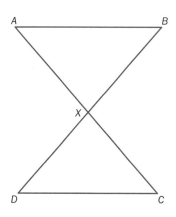

Show that *ABX* is congruent to *CDX*. (*3 marks*)

24

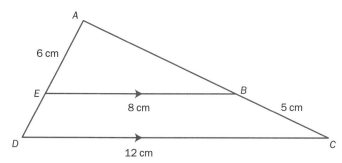

a Find the length of *AD*. (*2 marks*)

b Find the length of *AB*. (*3 marks*)

25 a Angle *AOC* = 110°

A, B and C are points on a circle, with centre O.

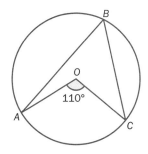

Diagram **NOT** accurately drawn

Find angle *ABC* Give a reason for your answer. (*3 marks*)

Answers

Exercise N1

1 1×120, 2×60, 3×40, 4×30, 5×24, 6×20, 8×15, 10×12

2 315 seconds

3 **a** 45 330
b 45 300
c 45 000

4 13 cm × 17 cm × 19 cm

5 4450 feet, 4350 feet

6 **b** Not prime, it is divisible by 2.
c Not prime, it is divisible by 3.
d Not prime, it is divisible by 7.
e Prime

7 **a** 65 s, 55 s
b 5 minutes 25 seconds, 4 minutes 35 seconds

8 **a** **i** $2 \times 3 \times 5$
ii 3×5^2
iii $2^4 \times 3$
b **i** 15
ii 6
iii 3

9 **a** **i** $2^2 \times 3^2$
ii 2^6
iii $2^4 \times 5$
b **i** 576
ii 320
iii 720

10 Yes, 65 kg + 75 kg + 85 kg + 85 kg = 310 kg > 300 kg

11 **a** $\dfrac{40 \times 400}{50} = 320$

b $\dfrac{200}{20 \times 100} = 0.1$

c $\dfrac{50 \times 50}{50 + 50} = 25$

12 **a** $3^2 \times 5 \times 7$
b 35

13 $m = 3$, $n = 5$

14 **a** 663
b **i** 2.21
ii 0.013

Exercise N2

1 **a** 5 thousandths
b 2 tenths
c 6 hundredths
d 4 ten-thousandths

2 **a** $\dfrac{30}{72}, \dfrac{32}{72}, \dfrac{28}{72}, \dfrac{27}{72}, \dfrac{34}{72}$

b $\dfrac{3}{8}, \dfrac{7}{18}, \dfrac{5}{12}, \dfrac{4}{9}, \dfrac{17}{36}$

3 **a** **i** $\dfrac{2}{5}$
ii $\dfrac{2}{25}$
iii $\dfrac{9}{50}$
iv $\dfrac{3}{8}$
b **i** 40%
ii 8%
iii 18%
iv 37.5%

4 **a** **i** $\dfrac{4}{5}$
ii $\dfrac{1}{8}$
iii $\dfrac{17}{40}$
iv $\dfrac{7}{80}$
b **i** 0.8
ii 0.125
iii 0.425
iv 0.0875

5 **a** **i** 0.45
ii 0.095
iii 0.046
iv 0.037
b **i** 45%
ii 9.5%
iii 4.6%
iv 3.7%

6 **a** 54
b 0 cars 13.3%, 1 car 41.7%, 2 cars 30%, 3 or more cars 15%

7 **a** $\dfrac{7}{8}$
b 87.5%
c One

8 **a** $0.\dot{4}$, recurring
b 0.375, terminating
c $0.\dot{5}\dot{4}$, recurring
d 0.3125, terminating

9 42 857

10 **a** $\dfrac{3}{5}, \dfrac{31}{50}, \dfrac{5}{8}, 63\%$

b $\dfrac{9}{10}, 0.905, 93.5, \dfrac{19}{20}$

c $16\%, 16.5\%, \dfrac{1}{6}, 0.17$

d $\dfrac{5}{8}, 66\%, \dfrac{2}{3}, 0.67$

e $\dfrac{4}{5}, \dfrac{21}{25}, \dfrac{17}{20}, \dfrac{43}{50}$

11 **a** 20
b 3.54
c 5.4
d 108
e 0.01

12 a $\frac{5}{9}$

b $\frac{9}{11}$

c $\frac{56}{111}$

d $\frac{17}{30}$

13 a $0.\dot{2}\dot{7}.$

b $100x = 39.393939\ldots$
$x = 0.39393939\ldots$
$99x = 39$ (subtract)
$x = \frac{39}{99} = \frac{13}{33}$

Exercise N3

1 a 7011
b 47
c 133

2

2	−5	0
−3	−1	1
−2	3	−4

3 a -2×-3
b 2×-3

4 Sophie, it must be calculated as $(16 + 14) \div (10 + 5)$

5 a 36
b −8
c 20

6 a 250
b 24
c 120

7 a $4\frac{1}{4}$

b $2\frac{23}{40}$

c $1\frac{13}{15}$

8 $3\frac{3}{5}$ cm^3

9 a $14\frac{2}{5}$

b $1\frac{1}{4}$

c $2\frac{2}{3}$

d $1\frac{1}{4}$

e $1\frac{1}{3}$

f 63

10 $\frac{17}{32}$ inch

11 $1\frac{3}{5}$ km

12 a 120 km/h
b Yes

13 a $\frac{1}{2}$

b $4\frac{7}{12}$

14 24 cm

Exercise N4

1 3.3

2 a 1960
b 280

3 a 10.35 m^2
b £15.53

4 a 48.4
b 300
c 20.7
d 0.91
e 0.010

5 a 2.171428571
b 2.2

6 a $3\frac{1}{4}$

b 3.682428682
c 3.7

7 a 75 minutes
b 129 minutes
c 220 minutes

8 a 13 500, 14 500
b 8450, 8550
c 12 500 13 500
d 7350, 7450
e 1850, 1950

9 a 495 km, 485 km
b 16.5 seconds, 15.5 seconds
c 3.005 m, 2.995 m
d 8.55 cm, 8.45 cm
e 25.05 seconds, 24.95 seconds

10 10.85 g/cm^3, 10 g/cm^3

11 a 100.5 mm
b 101.5 mm

12 17 bars

Exercise N5

1 a 1024, 729, 512, 625
b 2^9, 5^4, 3^6, 4^5

2 a 8^8
b 8^7
c 8^3
d 8^{-2}

3 a $\frac{1}{5}$

　b $\frac{1}{100}$

　c $\frac{1}{16}$

　d $\frac{1}{125}$

4 a 3

　b −1

　c 0

　d $\frac{1}{2}$

　e 27

　f 5

　g 81

　h 81

5 a 6

　b 2

　c 4

　d 16

6 a $6\sqrt{2}$

　b $3\sqrt{7}$

　c $6\sqrt{5}$

　d $8\sqrt{3}$

7 a $\sqrt{5}$

　b $4\sqrt{2}$

　c $6\sqrt{2}$

　d $\sqrt{3}$

8 a 6

　b $2\frac{1}{2}$

　c $\frac{\sqrt{5}}{2}$

　d $2\sqrt{3}$

9 a 59

　b $3 + 2\sqrt{2}$

　c $\sqrt{7}$

　d $\frac{\sqrt{5}-1}{4}$

10 a $6\sqrt{3}$ cm^3

　b $3\sqrt{3}$ cm^3

11 a i 5^6

　　ii 5^3

　b $x = 7, y = 3$

12 a 1

　b $\frac{1}{16}$

　c 6.35

13 $\sqrt{22}$

14 a 4

　b $k = 2$

　c 83.3%

Exercise N6

1 a 1:50 000

　b 1.6 km

　c 3 cm

2 £1 = ≅€1.49

3 a 36 km/h

　b $2\frac{1}{2}$ hours

4 a 6.68 m/s

　b 224.4 seconds

5 42 kg

6 £6.30

7 300 g

8 a 5:7

　b Emma £400, Ruth £560

9 a £320

　b Jim 50%, William 30%, Harry 20%

10 a 91.4 cm

　b 3.281 feet

　c 1 metre

11 a £21

　b 47 kg

Exercise N7

1 £5.25

2 2903 pixels × 2177 pixels

3 $7.98 \times 10^9 \, m^2$

4 a 3.3%

　b 50 bricks

5 42.9%

6 £50

7 a £1191.02

　b 7 years

8 a 0.85

　b £433.50

　c 5 years

9 £275

10 a £5062.50

　b 0.4096

11 a 4.5
   b £1205.86

Exercise N8

1

Power of 10	Meaning	Number
10^3	$10 \times 10 \times 10$	1000
10^2	10×10	100
10^1	10	10
10^0	1	1
10^{-1}	$\frac{1}{10}$	0.1
10^{-2}	$\frac{1}{10^2}$	0.01
10^{-3}	$\frac{1}{10^3}$	0.001
10^{-4}	$\frac{1}{10^4}$	0.0001

2 a 1×10^{-1}
   b 8×10^{-3}
   c 3.4×10^{-9}
   d 8.64×10^4
   e 3.1536×10^7

3 a 0.15, 0.105, 0.098, 0.089
   b 8.9×10^{-2}, 9.8×10^{-2}, 1.05×10^{-1}, 1.5×10^{-1}

4 Electron, proton, neutron

5 2326 clips

6 30 303 clips

7 a 240
   b 400 000
   c 0.000 42

8 1.55×10^2 cm

9 2.9×10^{-1} kg

10 a 8.467×10^9
   b 17.6%

11 a 2.1×10^{10} km^3
   b 1.1×10^{12} km^3
   c 1:53

12 1.9×10^7

Exercise A1

1 a $6a$
   b Length $6a$, width $3a$
   c $18a$

2 a $n + 4$, $2n + 8$, $n + 8$, $n - 2$, -2
   b -2

3 a $9 - 2x$
   b $2x - 22$
   c $2a^2 + 2b^2$

4 a $x(x - 8)$
   b $(x - 4)(x + 4)$
   c $(x - 2)(x - 8)$

5 $(2x + 1)(x + 4) = 72 \Rightarrow 2x^2 + 9x + 4 = 72$
   $\Rightarrow 2x^2 + 9x - 68 = 0$

6 a $12h^7$
   b $\frac{2b^3}{a}$
   c $25p^6$

7 a $\frac{x}{x + 1}$
   b $\frac{2x + 1}{x + 3}$
   c $\frac{x - 3}{x + 3}$

8 a $\frac{2x + 1}{x(x + 1)}$
   b $\frac{2x - 5}{(x - 2)(x - 3)}$
   c $\frac{12 - x}{(x - 2)(x + 3)}$

9 a $81x^4y^8$
   b $\frac{x}{x - 5}$

10 a a^7
   b $15x^3y^4$
   c $x - 1$
   d $(x + 3)(x - 3)$

Exercise A2

1 a $6a + 36 = 180$, $a = 24$
   b 74°, 106°, 74°, 106°

2 a $(6x + 6)$ cm^2
   b $x = 5$

3 a $2(3x - 1) = 8(x - 1)$, $x = 3$
   b 16

4 a $x \geqslant 2$
   b $x < 4$
   c 2, 3

5 a $x > -3$
   b

6 a $x \leqslant 4$ and $x > \frac{1}{2}$
   b 1, 2, 3, 4

7 a $x = 13$
   b $x = 4$
   c $x = 12$

8 $11 = 2m + c$, $23 = 5m + c$; $m = 4$, $c = 3$

9 $x = 1.5$, $y = 0.5$

10 $x = 4.5$, $y = -3$

11 a $x = 7$

   b $\dfrac{2x}{2x + 3}$

Exercise A3

1 a $x^2 = \dfrac{1}{x} + 1 \Rightarrow x^3 = 1 + x \Rightarrow x^3 - x = 1$

   b $x = 1.3$

2 a $(x + 3)(x + 8) = 0$, $x = -3$ or -8

   b $(2x - 1)(x + 3) = 0$, $x = \dfrac{1}{2}$ or -3

   c $(2x - 1)(3x - 1) = 0$, $x = \dfrac{1}{2}$ or $\dfrac{1}{3}$

3 a $x^2 + (x + 2)^2 = (x + 4)^2$ by Pythagoras $\Rightarrow$ $x^2 - 4x - 12 = 0$

   b $x = 6$ gives side lengths 6, 8, 10

4 $x = 1.18$ or -0.425

5 $x = 1.5$ or 0.5

6 a $x^2 + (x - 1)^2 = 25 \Rightarrow 2x^2 - 2x - 24 = 0$ $\Rightarrow x^2 - x - 12 = 0$

   b $(4,3)$, $(-3,-4)$

7 a $(2x + 5)(3x - 2) + 2(3x - 2) = 25 \Rightarrow$ $6x^2 + 17x - 39 = 0$

   b i $x = 1\dfrac{1}{2}$ or $-4\dfrac{1}{3}$

     ii $8\,\text{cm}$

8 a $x^2(x + 1) = 230 \Rightarrow x^3 + x^2 = 230$

   b $x = 5.8$

Exercise A4

1 a $\pi r^2 h$

   b $375\pi\,\text{cm}^3$

2 $A = (2r)^2 - \pi r^2 \Rightarrow A = r^2(4 - \pi)$

3 $r = \sqrt{\left(\dfrac{A}{\pi}\right)}$

4 a $a = \dfrac{2A}{h} - b$

   b $8\,\text{cm}$

5 a $x = \dfrac{b - a}{3}$

   b $x = \dfrac{b - a}{c - d}$

   c $x = \dfrac{kb - a}{1 - k}$

6 a $a = \dfrac{v^2 - u^2}{2s}$

   b 2

7 a $y = kx$, $y = \dfrac{k}{x}$, $y = kx^2$

   b C, A, B

8 $y = 90$ should be $y = 96$

9 a $d = 5t^2$

   b 245

   c 3

10 a $s = \dfrac{8000}{f^2}$

   b 500

Exercise A5

1 a 10, 100, 1000, 10 000, 100 000

   b 2, 4, 8, 16, 32

   c 8, 15, 24, 35, 48

2 a 9, 14, 19

   b $5n + 4$

   c The first cube has 4 edges at the front and 5 more visible edges. Each new cube brings 5 more visible edges.

   d $v = 5n + 4$

   e 20

3 a nth term is $n^2 + n(n + 1) - 3$

   b $(2n + 3)(n - 1) = 2n^2 + n - 3$ and $n^2 + n(n + 1) - 3 = 2n^2 + n - 3$ so $(2n + 3)(n - 1) = n^2 + n(n + 1) - 3$

4 For example, $n = 2$ gives $p = 9$ which is not prime.

5 For example, $0.1^2 = 0.01$ which is less than 0.1.

6 $\dfrac{n + 1}{2n + 3}$

7 a $r + 1$

   b $r^2 + (r + 1)^2 = 2r^2 + 2r + 1 = 2(r^2 + r) + 1$ which is one more than an even number so it must be odd.

8 a $2r + 3$

   b $(2r + 1)(2r + 3) = 4r^2 + 8r + 3 = 4(r^2 + 2r) + 3$ which is 3 more than a multiple of 4.

9 $d = 4n + 6$

10 a $(2a - 1)^2 - (2b - 1)^2 = 4a^2 - 4a + 4b^2 + 4b$ and $4(a - b)(a + b - 1) = 4a^2 - 4a + 4b^2 + 4b$ so $(2a - 1)^2 - (2b - 1)^2 = 4(a - b)(a + b - 1)$

   b $(2a - 1)^2 - (2b - 1)^2 = 4(a - b)(a + b - 1)$ by part **a**.
If $(a - b)$ is even then $(a - b)(a + b - 1)$ is even.
If $(a - b)$ is odd then a is odd and b is even or vice versa, so $(a + b)$ is odd and $(a + b - 1)$ is even. Again $(a - b)(a + b - 1)$ is even.
So $4(a - b)(a + b - 1)$ is 4 times an even number which is a multiple of 8.

Exercise A6

1 a

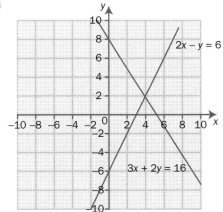

b (4,2)

2 a $y = -0.6x + 3$
  b i -0.6
    ii (0,3)

3 $2x + y = 4$, $a = 2$, $b = 1$, $c = 4$

4 a

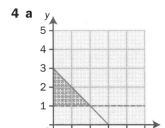

b (0,2), (0,3), (1,2)

5 $y = -2x + 5$

6 a 4
  b $-\frac{1}{4}$
  c (2,5)
  d $y = -\frac{1}{4}x + 5\frac{1}{2}$

7 a -1, 0, 1
  b $(-1,-1)$, $(0,-1)$, $(1,-1)$, $(0,0)$, $(1,0)$, $(1,1)$

Exercise A7

1 P $(-2,-4)$, Q $(3,6)$

2 a $(x + 2)(x - 3)$
  b A $(-2,0)$, B $(3,0)$
  c $x = 0.5$

3 a

x	-3	-2	-1	0	1	2	3
y	0.125	0.25	0.5	1	2	4	8

b

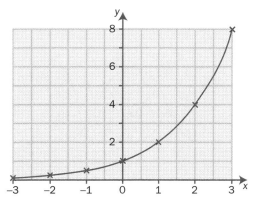

c i $x = 1.6$
  ii $x = 2$, $x = 1$

4 a $x^2 + (x + 3)^2 = 9 \Rightarrow x^2 + 3x = 0$
  b (0,3), $(-3,0)$

5 a

x	-2	-1	0	1	2
y	-12	-4	-2	0	8

b

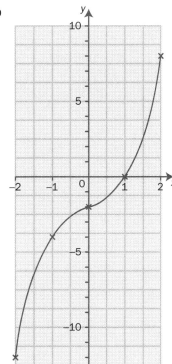

6 a i (90,1)
    ii (180,0)
  b i (45,0)
    ii (54.7,-1)

Exercise A8

1 A (0,0.5), B (6,5), C (12,0.5)

2 B

3

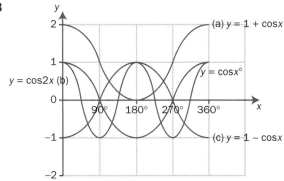

4 a

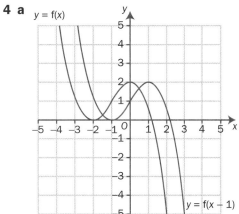

b

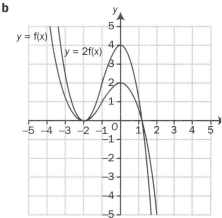

5 a $p = 1600$, $q = 0.5$
 b £6400

Exercise S1

1 125°, 120°, 51°, 64°

2 a $a = 70°$
 b $b = 105°$, $c = 60°$

3 a 15°
 b 165°

4 24°, 48°, 72°, 96°, 120°

5 a 14
 b 15

6 a 220°
 b 040°

7 a 720°
 b 95°, 130°, 140°, 120°, 130°, 105°

8 56°

9 a 30.7°
 b 121°

Exercise S2

1 a $a = 30°$, $b = 35°$, $c = 65°$, $d = 50°$

 b

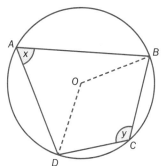

Obtuse angle $DOB = 2x$ (The angle at the centre of a circle is double the angle at the circumference from the same arc.)
Reflex angle $DOB = 2y$ (The angle at the centre of a circle is double the angle at the circumference from the same arc.)
Obtuse angle DOB + reflex angle $DOB = 2x + 2y = 360°$ (Angles at a point sum to 360°.)
So $x + y = 180°$

2 $x = 72°$, $y = 108°$, $z = 36°$

3 $8^2 + 15^2 = 17^2$, so by Pythagoras' theorem ABC is a right-angled triangle with right angle at A. As angle A is 90° it must be the angle in a semicircle, which means CB is a diameter of the circle, so it passes through the centre.

4 $a = 50°$, $b = 65°$

5 Both triangles are right-angled, the hypotenuses are equal ($OA = OB$ as they are both radii) and both triangles share OM so triangles OAM and OBM are congruent by RHS.

6 Angle $ABC = 90°$ (angles in a semicircle) and angle $ABD = 90°$ (angles in a semicircle), so angle $CBD = 180°$ which means CBD is a straight line.

7 a i 150°
 ii The angle at the centre of a circle is double the angle at the circumference from the same arc.
 b 30°

8 a 60°
 b 35°
 c Yes, angle $DBA = 90°$ so it is the angle in a semicircle with BD as diameter.

Exercise S3

1 Parallelogram

2 a Triangular prism

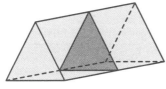

b There are 4 planes of symmetry.

3 a (−2,−1), (2,−3)
   b Both are 7 square units.

4 a Square, rhombus, kite
   b $2x^2$

5

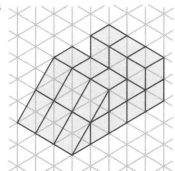

6

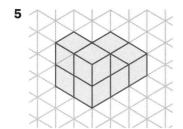

7 a

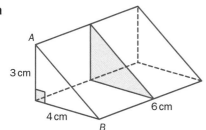

   b 5 cm
   c

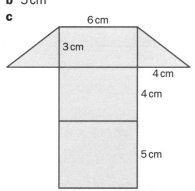

d 84cm^2
e 36cm^3

8 A (2,2,4), B (2,2,0), M (2,2,2)

Exercise S4

1 a 90°
   b 20 cm
   c Area of circle − area of triangle = $100\pi - 96 =$ $4(25\pi - 24)\text{cm}^2$

2 2.5 cm

3 0.5m^2

4 a $14\pi r^2 + 4\pi rh$
   b $3\pi r^2 h$

5 Area of sector = $\frac{30}{360} \times 36\pi \text{cm}^2 = 3\pi \text{cm}^2$, area of triangle = $\frac{1}{2} \times 12 \sin 15° \times 6 \cos 15° = 9 \text{cm}^2$, shaded area = $3\pi - 9 = 3(\pi - 3)\text{cm}^2$

6 53.8cm^2

7 $(18 + 2\pi)$ cm

8 $h = 32x$

Exercise S5

1 a Half turn about $\left(-1\frac{1}{2}, 1\right)$
   b Reflection in the line $x = 1$
   c Translation $\begin{pmatrix} 3 \\ -5 \end{pmatrix}$
   d Reflection in the line $y = 1$
   e Half-turn about (1, 1)

2 a,b

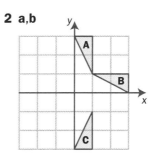

   c Reflection in the x-axis

3 a 2a
   b a + b
   c a − b
   d 2a − b

4 a b − a
   b a
   c a + b
   d $\frac{1}{2}$b
   e $\frac{1}{2}$b
   $\overrightarrow{MX} = \overrightarrow{XN}$ so MX and XN are parallel. MX and XN also have the point X in common so M, X and N are on the same straight line.

5 Reflection in $y = x$

6 a 2a – 2b

  b $\overrightarrow{XY}$ = 2a which is a multiple of $\overrightarrow{OR}$ so XY is parallel to OR.

Exercise S6

1 $\frac{420}{297}$ = 1.414 141 414, $\frac{297}{210}$ = 1.414 285 714, similar to 3 d.p.

2 a Angle ACB = Angle ECD (vertically opposite angles), angle CAB = angle CDE (alternate angles) and angle ABC = angle DEC (alternate angles) so triangle ABC is similar to triangle DEC.

  b CB = 8.8 cm, CD = 10 cm

3 a

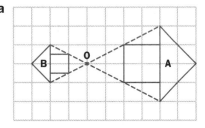

  b Enlargement scale factor –2, centre at origin

4 a 15.625 : 1

  b 6.25 : 1

  c 6 cm

5 62.5 litres

6

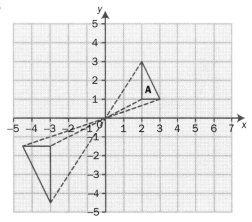

7 a 12 cm

  b 2700π cm³

Exercise S7

1 a Tetrahedron

  b Accurate net of tetrahedron

2

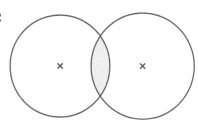

3

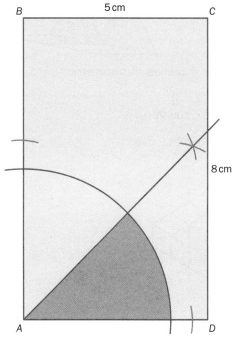

4 a Either of:

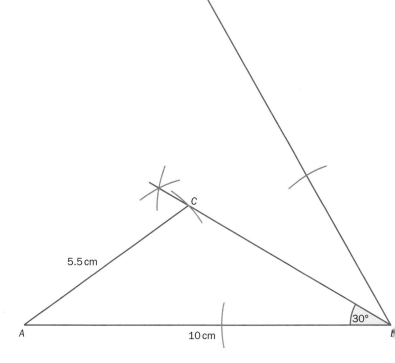

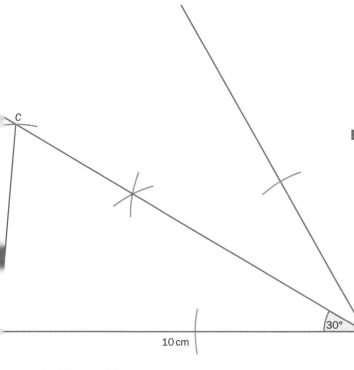

10 cm

30°

b 65° or 115° depending on triangle drawn
c No, there are two possible triangles.

5 a *AP* = *BP* (same arc length), *AQ* = *BQ* (same arc length) and *PQ* = *PQ* (side common to both triangles) so triangles *PAQ* and *PBQ* are congruent by SSS.
b *APBQ* is a rhombus (equal sides) so the diagonals bisect each other at right angles. As *M* is the point where the diagonals cross, it is the midpoint of *AB*.

6

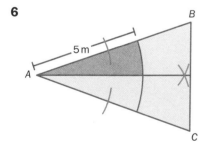

5 m

7 a *AB* = *AC* (triangle *ABC* is isosceles), *BP* = *CP* (two tangents to a circle from an external point are equal in length) and *AP* = *AP* (side common to both triangles) so triangles *APB* and *APC* are congruent by SSS.
b 50°

Exercise S8
1 6.6 cm

2 a 138°
  b 318°

3 (1, 7)

4 a 14.1 cm
  b 7.07 cm
  c 54.7°

5 26.4°

6 a 34.8°
  b 15.4 cm

7 22.2°

8 18.3 cm

Exercise D1
1 a

Time (sec)	Frequency
20 < t ⩽ 25	3
25 < t ⩽ 30	8
30 < t ⩽ 35	16
35 < t ⩽ 40	11
40 < t ⩽ 45	2

b 30 < t ⩽ 35

2 a

	Men	Women	Total
20 ⩽ a < 40	1	7	8
40 ⩽ a < 60	1	15	16
Total	2	22	24

b 33.3%

3 a Only students whose surnames begin with letters at the start of the alphabet will be included in the sample.
  b For example, give each student a number and use random number tables to pick numbers.
  c Increase the sample size by including more than one student from each form.

4 a His survey will only include people who use this particular hotel. It will also only include people who stayed on a weekend night (Sunday) rather than a week night.
  b It is a leading question, suggesting that the correct answer is 'Yes'.

5 If the machine was not producing faulty clips at random, the systematic sample could be biased. For example, if every 100th clip is the wrong shape, sampling the 100th, 200th, etc. would pick every faulty clip whereas sampling the 101st, 201st, etc. would pick none of them.

6 For example:
  How many hours did you sleep last night?
  Less than 5 hours
  5 or more hours but less than 6 hours
  6 or more hours by less than 7 hours
  7 or more hours but less than 8 hours
  8 or more hours

7 a 24.02 km/h, 23.98 km/h
  b The upper and lower bounds are the same to 1 d.p. so 24.0 km/h is an appropriate value.
  c 18, 12, 9, 9, 2

Exercise D2

1

4	5 7 8 9
5	0 0 1 2 3 4 6 7 7 7 8 8 8 9
6	0 0

Key: 4|5 means 4.5

2 Pie chart with: Mayonnaise 120°, Nuts 114°, Peanut butter 75°, Crisps 51°

3 a,c

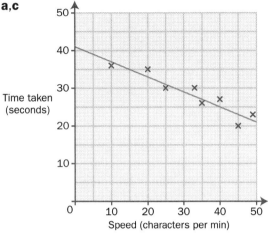

Time taken (seconds) vs Speed (characters per min)

b Negative
d 29 seconds
e A speed of 70 characters per minute is outside the range of data collected.

4 Pie chart with: England 5°, Scotland 337°, Wales 10°, Ireland 8°

5 a,c

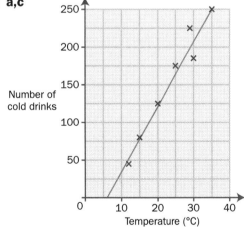

Number of cold drinks vs Temperature (°C)

b Positive
d 158
e 40° is outside the range of data.

6

0	5 7 8 8
1	0 0 0 0 2 5 5 5 6
2	0 0 0 4 5
3	3 5

Key: 1|5 means 15

7 Pie chart with: Gold 60°, Silver 80°, Bronze 220°

Exercise D3

1 a There should be no gaps between the bars. The class intervals should not overlap. The bars should be different widths because the class intervals are different sizes. The area of the bar should be used to represent the frequency not the height.

b Class width: 5, 5, 10, 20, 20; Frequency density: 1.2, 3, 3, 1.75, 0.7

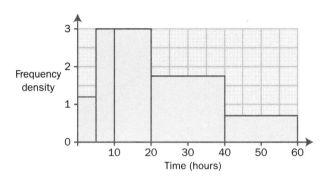

Frequency density vs Time (hours)

2 20, 18, 45, 52

3 a Cumulative frequency:
10, 25, 55, 90, 115, 120

b

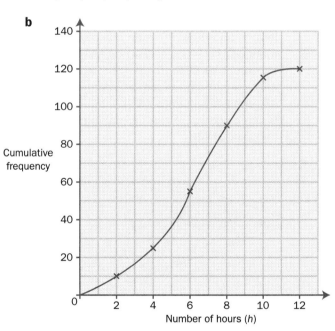

Cumulative frequency vs Number of hours (h)

c 73 children

Exercise D4

1 a 19 lessons
b Median = 18 lessons, UQ = 24 lessons, LQ = 13 lessons
c Range = 29 lessons, IQR = 11 lessons
d 18 lessons
e

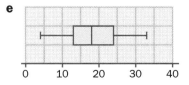

2 a i 51 bedrooms
  ii 20 houses
  iii 2.55 bedrooms
 b 2 bedrooms

3 a £15 < v ⩽ £20
 b £10 < v ⩽ £15
 c £13.10

4 3.72 minutes or 3 minutes 43 seconds

5 a £37, £45, £48, £51, £50, £49, £43.50, £39.50, £32.50
 b September to December

6 a i 152 cm
  ii 177 cm
 b

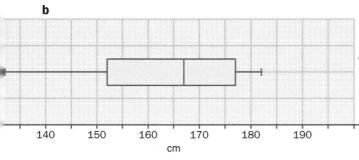

7 a 182, 178, 180, 184
 b 0.8 × 0.7 = 0.56 so it is a 44% reduction.

Exercise D5

1 The modal weight is higher for the girls. The boys' weights are more spread out.

2 a Alan: 15, 40, 78, 96, 100, 100;
  Zena: 10, 24, 40, 64, 90, 100
 b

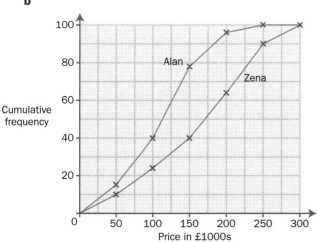

 c Alan: median = £113 000, IQR = £76 000;
  Zena: median = £170 000, IQR = £119 000
 d On average, houses are more expensive in Zena's neighbourhood (higher median). House prices are also more variable in Zena's neighbourhood (higher IQR).

3 a,c

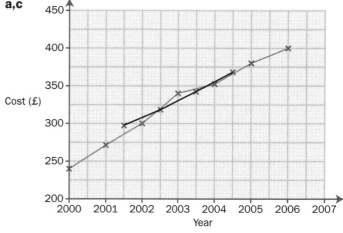

 b £288, £316, £343, £368
 d About £393
 e About £440

4 a i Electricity £10, gas £19
  ii Electricity £5, gas £8
  iii Electricity £60, gas £59
 b The gas bills were lower on average (lower median). The gas bills are more variable than the electricity bills (higher range and interquartile range).

5 The range is the same for both types of rose plants (20 cm). The modal class for the miniature rose plants is 35–40 cm, the modal class for the standard rose plants is 65–75 cm, so the standard rose plants are taller on average.

Exercise D6

1 a

	Flat	Town house	Semi-detached	Total
Male	19	15	16	50
Female	7	9	14	30
Total	26	24	30	80

 b $\frac{13}{40}$

 c $\frac{5}{8}$

2 a 0.2, 0.15, 0.17, 0.15, 0.16
 b

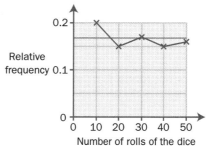

 c 0.167

3 a

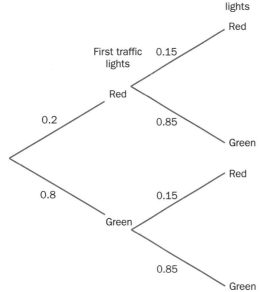

First traffic lights
Second traffic lights

Red — 0.15 — Red

Red — 0.85 — Green

0.2

0.8 — Green — 0.15 — Red

0.85 — Green

b 0.68

4 a 0.2
b 60

5 a $\frac{8}{25}$
b £4

6 $\frac{33}{50}$

Higher Practice Exam Paper Answers

1 a 12°C
b −1°C

2 a i 5:7:8
ii 100:140:160
b i 5:3
ii $\frac{3}{8}$

3 a 5^8
b $3^2 \times 5$

4 6.25 cm²

5 a $A = (x - 3)(2x - 1) = 2x^2 - 6x - x + 3$
$= 2x^2 - 7x + 3$
b $P = 6x - 8 = 28$, therefore $x = 6$ and $A = 33$

6 a 14
b $2 + 3n$
c No, 3 does not divide into 245.

7 a

x	−3	−2	−1	0	1	2	3
y	11	9	7	5	3	1	−1

b

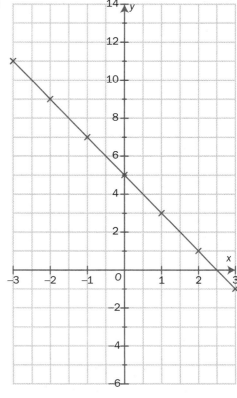

8 a $y = 17$
  b B
  c 5
  d $y = 4x + 2$ (any equation of the form
    $y = 4x + c$ for $c \neq 2$ will do)
  e 1

9 a

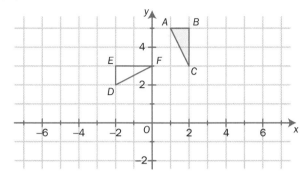

  b rotation 90° clockwise about (1,2)

10 Total area = 144 + 28 = 172 cm²

11 a 16
  b The people from each group have not been randomly selected for the sample.

12 a

1	1 2 6
2	3 7
3	1 4 5
4	5 8 9
5	1 1 2 3 6 7
6	1 2 3

Key: | 1 | 1 | means 11 years old

  b 48
  c

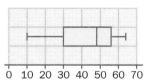

13 a

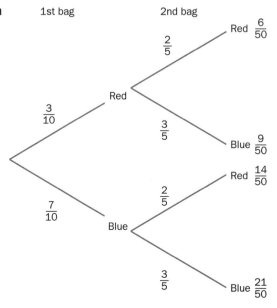

b $\frac{29}{50}$

14 a Negative
  b

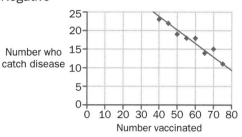

  c 19

15 99.58 kg

16 a No
  b $x = 10$
  c $y = 2x + 3$ (Any equation of the form $y = 2x + c$ where $c \neq -1$ will do.)

17 $x = 3$, $y = -1$

18 a $7(2x + 3) - 3(x - 4) = 14x + 21 - 3x + 12$
    $= 11x + 33$
  b $(2x + y)(x + 3y) = 2x^2 + xy + 6xy + 3y^2$
    $= 2x^2 + 7xy + 3y^2$

19 a $10x^2 - 15xy = 5x(2x - 3y)$
  b $p^2 - 49q^2 = (p + 7q)(p - 7q)$
  c $3x^2 + 7x - 6 = (3x - 2)(x + 3)$

20 a $x = -1$ or 6
  b 2.351 or −0.851

21 a ADE = ABC. ABC = 180 − 105 = 75°, therefore ADE = 75°
  b AED = 180 − (ADE + EAD) = 180 − (75 + 43)
    = 180 − 118 = 62°

22 8 sides (Octagon)

23 Since X is the midpoint of AC, AX = XC. Since AB and DC are parallel, BX = XD, so AB = DC. Also, ABX = CDX and BAX = DCX, therefore they are congruent.

24 a AD = 9 cm
  b AB = 10 cm

25 ABC = 55° since the angle at the centre of a circle is exactly double that of the angle at the circumference from the same arc.